THE SEEDS OF MODERN SCIENCE

germinated about 4000 B.C. in Mesopotamia—in a Sumerian civilization that had evolved the wheel, a calendar, and a form of writing and arithmetic. But it was not until the advent of the early Greeks—men such as Thales and Pythagoras, who conducted a rational inquiry into the nature of the universe—that science became disassociated from religion, myth, and magic.

This brilliant and informative account follows the progress of science throughout the centuries. The authors chronicle the enlightened period of Plato and Aristotle . . . the great Hellenistic age of Euclid and Archimedes . . . the decline into darkness during the early Christian period—a period that lasted until the Islamic invaders brought to Europe the vast treasures of Greek and Arabic learning.

Discussing the work of the great men of medicine, Hippocrates, Galen, Harvey; and of astronomy, Ptolemy, Galileo, Newton, the Halls trace the course of science from ancient times to the twentieth century, showing how the Renaissance and the Industrial Revolution caused a forward surge in intellectual endeavor and how the nuclear physics of today has opened new worlds to mankind.

Other SIGNET SCIENCE LIBRARY Books

A BRIEF HISTORY OF SCIENCE

by A. RUPERT HALL
and MARIE BOAS HALL

A SIGNET SCIENCE LIBRARY BOOK
PUBLISHED BY THE NEW AMERICAN LIBRARY

FIRST PRINTING, SEPTEMBER, 1964

SIGNET TRADEMARK REG. U.S. PAT. OFF. AND FOREIGN COUNTRIES
REGISTERED TRADEMARK—MARCA REGISTRADA
HECHO EN CHICAGO, U.S.A.

SIGNET SCIENCE LIBRARY BOOKS are published by The New American Library of World Literature, Inc. 501 Madison Avenue, New York, New York 10022

PRINTED IN THE UNITED STATES OF AMERICA

PREFACE

If this book were planned according to the volume of scientific discovery in different periods, everything before 1800 would have to be summarized on the first page. To proceed thus would be to ignore the importance of the gradual development through time of scientific methods and concepts, the tracing of which has been our chief concern. For this reason we have devoted two-thirds of the book to science before the nineteenth century; the last third deals with modern science and its great accomplishments, theoretical and practical. In this final section it was impossible to describe all scientific developments, even the major ones, and we have therefore preferred to dwell upon a small number of topics of high importance and intrinsic interest.

In the earlier chapters we have employed the familiar forms of proper names (usually the Latin ones) but for the sake of distinction we have retained the Greek termination in *-os* where it is appropriate. Thus we write Alhazen, not Ibn al-Haitham; Epicuros, not Epicurus. Similarly we have omitted the diacritical marks used by scholars in making transliterations.

A short list of readings in which the matters discussed in each chapter may be further explored will be found at the end of the book; the sources of quotations are also indicated. In some cases a good, modern treatment in English still does not exist, and this is especially true of the modern period.

Our debt to fellow historians of science and to colleagues in more than one university is too great to be acknowledged in detail. We hope that all who merit our thanks will accept them. We also owe much to past and present students; to one of these, Mr. Victor E. Thoren, we are grateful for his care in reviewing the whole manuscript.

A. RUPERT HALL

MARIE BOAS HALL

Indiana University,
15 December 1962

CONTENTS

I

Philosophy and Physics in the Ancient World

Chapter 1

THE OLDEST CIVILIZATIONS

Man has always sought to command nature. Gradually he tried to understand nature. Much later he learned to combine the two desires. When that happened modern science took shape, but the development of modern science rests upon the curiosity and interest of many centuries, in which techniques for exploring nature were slowly developed and knowledge about nature slowly accumulated. The roots of modern science lie deep in the past and there is no one instant of time of which we can say, "Look, here science truly begins." At every stage of development there are both relics of the past and intimations of the future. And so the history of modern science truly begins at the beginning of things, though its first beginnings are faint indeed.

All men have found means to exercise control over nature. The most primitive human beings shaped stones into tools and used tree branches as levers. Most groups of men soon learned to control fire, and thereby not only began to make themselves to some small degree independent of the climate (as animal skins helped them to do too, of course) but to alter natural objects. The cooking of food is but a stage on the road to the smelting of ores, the baking of clay into bricks and pots, and the making of crude glazes and glass. Similarly the beginnings of agriculture and the domestication of plants and animals mark further important stages towards making man apparently the master of his environment.

Primitive technology was not the only means whereby man sought to subdue nature: he also tried magic and myth. Both have sometimes been spoken of as pre-science, but they are perhaps better regarded as theoretical technology. For the purpose of magic is to alter the course of nature: to cure the sick who would otherwise die, to prevent naturally ordained disaster from striking, or to cause the naturally strong and healthy to sicken and die. So too cosmogonic myths—the startlingly similar stories of the creation of the earth, sun, moon, and stars which are found throughout the world—are intended to help man walk unharmed through the mysteries of life and death. And though magic recognizes certain immutable laws—certain actions will always produce certain results, like cures like, what is once a part of an object is always mysteriously connected with it—it does not produce science because its aim is strictly utilitarian. Magic and myth are predecessors of science but can never become science.

The true origins of modern science lie in certain techniques developed in two seminal Mediterranean cultures, one in Mesopotamia, the land between the Tigris and Euphrates rivers, the other along the Nile. The Mesopotamian civilization is marginally the older. Here, in the south, there appeared about 4000 B.C. (about the time, curiously, of the creation of the world according to the reckoning of seventeenth-century historians) a people known to history as the Sumerians. There were men living precariously in flood-ravaged villages (the flood was Noah's) whom the Sumerians subdued, partly by their relatively advanced technology. They understood irrigation, they lived in cities and practiced extensive commerce, and they possessed the wheel—equally useful on war chariots and in the form of the potter's wheel. As a commercial people they were interested in numbers and in designating private possessions, and their writing is thought to have derived from the carved cylinder-seals used to denote ownership. It soon took the aspect known as cuneiform, the result of writing on soft clay by incising marks with a wedge-shaped stylus. Cuneiform, like Japanese writing, uses a syllabic, rather than an alphabetic, system; as the centuries went on, and the language changed, "dictionaries" were compiled—lists of symbols with ancient and later equivalents—and stored in libraries after baking. Perhaps a thousand years after the southern Sumerians arrived they were conquered by the Akkadians of the north; several hundred years later further conquerors established a new kingdom with its capital at Babylon. The Babylonian Empire slowly merged after 1000 B.C. with the Assyrian Empire, and this in turn, after 500 B.C., became successively part of the Persian, Greek, and Roman empires. But its culture remained individual until at least 300 B.C., and perhaps even later.

The Sumerians, and after them the Babylonians, were not significantly scientific in outlook as far as we know, but they made major contributions to later science through their mathematical and astronomical techniques. Although cuneiform does not allow ease of arithmetical reckoning, the Sumerian scribes gradually developed a simple but effective technique which is, mathematically speaking, sophisticated. Their numbers were, like their syllables, marked on clay with a stylus; and they found that two symbols alone would suffice: a vertical wedge represented the number 1 and a horizontal wedge the number 10. With these two, any number could be represented. The Sumerian was not a decimal system like ours, but sexagesimal, that is, based on the number 60; it was otherwise rather like our familiar system, in that it made use of *position*. Just as each "3" in the number 333 represents a different value, depending on its place (the right-hand digit three, the middle digit thirty and the left-hand digit three hundred), so the position of a wedge indicated whether it was a digit or sixty times that digit or sixty squared (3,600) times that digit, and so on. What the Sumerian or Babylonian system lacked was a zero, and so they had nothing to mark an empty space until very late indeed in their history (after 500 B.C.). This handy and ingenious system was excellent for representing large numbers; it was perhaps less convenient for quick calculation, and in fact was used for mathematics only, not in everyday life. To aid in calculation the Babylonians drew up tables, analogous to their earlier lists of syllables. They not only had multiplication tables of great complexity; they had tables of squares and cubes, square and cube roots, reciprocals, and even tables providing the solution to what we should now represent by linear and simple quadratic equations. And by about 1800 B.C. their ability to solve algebraic and geometrical problems became very great. They could solve fairly complex equations, but they did so always in numerical terms, and always by concrete example, for they did not possess the notion of generality. Centuries later their results were used by Greek mathematicians who had achieved general methods, and so indirectly the Babylonians influenced the development of modern mathematics. In the same way their use of a sexagesimal system has bequeathed to us, through astronomy, the sixty seconds in a minute, sixty minutes in an hour, and the three hundred and sixty degrees in a circle.

Because we possess only mathematical tables and tablets setting out problems, we know little about the thought of the Babylonian computers and mathematicians. But nothing leads us to believe that they had any theoretical interest whatsoever. Assessment is even more difficult when it comes to Babylonian work in astronomy. For here too there exists a mass of data with little indication of any underlying theoretical framework, and there is even more discontinuity between the earlier and later periods that exists in the case of mathematics. About Sumerian interest in the heavens we know nothing except that they used a lunar calendar—suitable for an agricultural people, but annoying in more complex forms of life since lunar months vary in length, and there is not a whole number of lunar months in a solar year. By the Babylonian period (after 2000 B.C.) we know more: many observations were recorded of the motion of the moon, particularly its rising and setting positions, and there were records of the rising and setting of Venus and Mercury, the two planets which keep close to the Sun, and are therefore observed near the horizon. Other horizon phenomena were recorded as well; atmospheric phenomena like haloes and peculiar clouds are listed side by side with eclipses and strange planetary configurations. It was in this period that the constellations were named, and the heavens divided into three zones, each divided into twelve sections. Out of this primitive system of mapping developed the zodiac (as the Greeks named it), the celestial band extending on either side of the equator in which the Sun and all the planets are to be found, which was divided into twelve signs, houses, or constellations. The purpose of all these celestial records was astrological; it was thought that in the heavens could be found signs and omens indicating the future welfare of the state. (Only later did "Chaldean astrology," as the Romans knew it, develop, with its insistence that the fate of each individual was written in the stars.) As the stars were named after and often called gods, and each city or state had its protective deities, these astrological preoccupations were furnished with a peculiar logic.

Very much later—well after 500 B.C., when Mesopotamia was no longer an independent, isolated state, but a part of the Greek-dominated Near East—a highly elaborate and complicated mathematical astronomy developed. It was primarily concerned with the very complex problem of lunar theory and to a certain extent was related to calendrical computation. In fact, after 300 B.C. two calculations of the length of the solar year were made which are accurate to within a very few minutes; it is significant that these are associated with the names of men to whom can be ascribed tables of lunar motion, with mathematical analyses of the effect of various sorts of variations which go to make up the observed differences in elapsed time between new moons. Modern mathematicians rightly admire the ingenuity, but astronomers find little relation to the science of astronomy as they know it. Notably, the Babylonians made no attempt to explain physically the movements that they computed.

The situation in the relatively nearby civilization of Egypt was comparable to that of the Mesopotamian basin. In Egypt agriculture was simpler than in Mesopotamia; men waited until the Nile rose and flooded its banks (or, occasionally, disastrously failed to do so), as it mysteriously did about the same time every year, and in the rich alluvial mud left behind after it subsided they planted crops. There was no problem of irrigation, and normally no problem of food supply. Egypt developed through the millennia a rich and cultured society of priests and aristocrats supported by an overwhelmingly larger population of peasants and artisans. Egypt, like Babylonia, was strongly theocratic, and learning was in the hands of the priests in both cultures. A surplus of labor, combined with an exceptionally complex cult of the dead, created the elaborate tombs and monuments familiar to us as pyramids and obelisks. The colossal size and careful workmanship of these great structures suggest to the modern eye a complex technology. In fact they were built with wedges and stone hammers to split the rock, sledges and ropes to drag the stones to the building sites, ramps from one level to another up which successive courses were hauled, levers to propel the stones into place, and water used to check when all was level. The Egyptians had no wheels or pulleys in the Pyramid Age (from about 2700 to 2000 B.C.), and the secret of their success was unlimited manpower, patience, and a strong artistic sense.

Interestingly connected with the pyramids is one of the oldest medical texts which we possess, the Edwin Smith papyrus. This is a copy (made about 1700 B.C.) of part of a treatise ascribed to Imhotep, originally a real person, an official at the court of King Zoser who built the first step pyramid and was later transformed into the god of medicine.

The scribe who wrote our copy obligingly included a glossary, for many technical terms had become obsolete in the course of a thousand years. The Edwin Smith papyrus deals with the treatment of injuries to the head and chest; the rest of the treatise is lost, though as a curious appendix there are some cosmetic recipes. The treatise follows a systematic method, describing first the examination of the patient, then the prognosis (a fatal injury being described as "an injury not to be treated"—though in fact treatment is always prescribed) and finally the treatment. Examination of wounds was highly skilled, and many obscure consequences of head injuries were known. Treatment consisted of bandages, application of medicaments, immobilization of the affected parts, and nursing. The frequency of fractures and head injuries suggests accidents in building rather than war injuries; this is a very early example of industrial medicine. It is of course purely empirical and technical; there is no indication that the surgeon who compiled the original text knew anything of the workings of the human body. Other texts show that there was, as one would expect, a strong magical element in the medicines prescribed for wounds and for internal consumption, though in historical times material remedies were preferred to incantation.

Egyptian mathematics has retained a certain historical fame, because the Greeks always insisted that their ancestors had learned geometry from the Egyptians. But in fact it was very elementary and, like the Babylonian or early Chinese methods, Egyptian methods were crude and empirical, developed for problem solving and not for theoretical purposes. Egyptian problems deal with acreages, or with measurements of stones for building, or are of the form "if so many men can do so much work in a certain number of hours, how much work can some other number of men do? . . ." The Egyptian number system was cruder than the Babylonian, and their methods of calculation very crude indeed. The Egyptians had symbols for one, five, ten, and multiples of ten; all other numbers were made up by adding together the appropriate number of symbols, rather like the Roman method. Anyone who has ever contemplated the possibility knows that multiplication with Roman numerals is ridiculously difficult; the Egyptians escaped this difficulty by performing multiplication by means of addition. Suppose an Egyptian wished to multiply 3 by 6; he wrote down two columns, beginning with 1 in the left and 3 in the right. He then doubled each successive figure in each column until in the left-hand column he found digits which would add up to 6 (2 plus 4); the corresponding numbers in the right hand column are 6 plus 12 or 18. The same process was used to divide 18 by 3. This method can be and was adapted for fractions, which were almost always those with the numerator 1; thus ¾ might be represented as ½, ¼. It was all time-consuming, but easy to learn, and it sufficed for the uses to which it was put. For, unlike the Babylonians, the Egyptians showed little interest either in astronomical events for astrological reasons, or in mathematical predictions of the motions of the moon. They used both a lunar and a solar calendar, and may have compared the two to determine the lapse of long cycles of time. But their reference point was the rising of certain notable stars, especially Sirius, not the erratic wanderings of Venus or the moon, and they never ventured into mathematical astronomy. Egyptian mathematics and calendrical astronomy, like Egyptian medicine, reached its peak early; it saw no complex growth nor late flowering as Babylonian mathematics and astronomy were to do. Its only legacy was, for no apparent reason, the 24-hour day.

To us, aware of the possibilities inherent in the techniques developed in Egypt, and still more in Babylonia, it is hard to realize that there was little of what we recognize as science or mathematics in the techniques so expertly practiced. Sophisticated as these often were, they remained at a purely technological level. The men who devised these methods were as skillful as one could find in any culture. But they lacked any spark of curiosity about why these techniques worked. Even in mathematics they never rose above the solution of problems, and so they remained on an elementary level. At no point did they begin to speculate about the nature of the world around them. We are by no means sure why science flourishes at one time and not at another, in one culture but not in another. But we do know that curiosity is an essential ingredient, and that until men begin to ask questions and to demand answers conceived in the same terms as the questions there will be no true science, for it is not science to say that a thing is so because a divine power chose to make it so. The Egyptians and Babylonians were deficient, not in technical skill and knowledge, but in curiosity. Desiring so little to understand the phenomena of the natural world, they influenced their successors technically but not conceptually.

Chapter 2

THE GREEK VIEW OF THE WORLD

In almost complete contrast to the Mesopotamian and Egyptian civilizations, the Greek civilization was one in which there were men who desired to know and understand, more than they desred to do. They were interested in ideas rather than techniques, and though they became possessed eventually of the techniques which earlier societies had developed in mathematics, astronomy, and medicine, they transmuted these techniques by the force of their notion of what is interesting in nature for a free man to know.

The early history of the Greeks is fairly obscure. Various tribes of Greek-speaking people began drifting into the region about the Aegean Sea after 2000 B.C.—at the beginning of the first great flowering of Babylonian mathematics and astronomy, and at the end of the Egyptian Pyramid Age. Exactly how they mixed and mingled with the peoples already there—including those of the advanced Minoan culture encountered on the island of Crete—is not clear, nor is it clear how they survived when Minoan civilization ended. Their life in the transition period is reflected in the Homeric poems; it is intellectually primitive, showing little interest even in astrology or calendrical astronomy. By the seventh century B.C., however, these warrior tribes had long since become by and large a city people, settled on the coastal strips of Asia Minor, Greece, South Italy, Sicily, and even sparsely in southern France and Spain. They had developed various new political forms, and had already a rich literary heritage. They had adopted a true alphabetic form of writing (traditionally from the Phoenicians), with a much more manageable and easily learned script than either cuneiform or hieroglyphic. At the same time they used their letters as numerals, in the same way as the Romans were to do; a cumbersome system, but one with which they were to remain quite content.

An unpromising beginning. Yet when the Greeks began to concern themselves with nature at all it was in a totally novel way. For they showed themselves interested in abstract ideas rather than in techniques, in natural philosophy rather than in computation or prediction of events. And this development was, like modern science, associated with individuals; it was not anonymous, like the development of all but recent technology.

At the beginning of the sixth century B.C.,* Thales, a successful merchant from the thriving commercial city of Miletus on the coast of Asia Minor, began a philosophic and scientific tradition which has never entirely died out. Thales left no writings, though he had pupils; what we do know about him besides a tradition of great wisdom is scanty though highly suggestive. He is said to have predicted an eclipse (probably that of 585 B.C.); to have been able to determine the distance from shore of a ship at sea; to have brought geometry from Egypt and to have discovered several geometrical theorems; and to have declared that the cosmos was made of water. His accomplishments could indeed, as later Greek historians supposed, have been acquired in the course of his travels—except his view of the cosmos. His declaration that the cosmos was made of water (or from water) implies a new and potentially revolutionary turn of mind. In the first place, Thales used the word *cosmos* for universe or world. When later Greeks spoke of the *cosmos,* they implied an ordered, rational, reasonable, comprehensible world, in which there was an explanation for all phenomena in natural (not supernatural) terms. Thales initiated the notions that to understand the cosmos one needed to know its nature (*physis,* hence the modern *physics*) and that this nature was to be interpreted as material. He was thus the founder of

* The Greeks cared little about exactitude in dating, and it is not until several centuries later that people began to have recorded birth and death dates. Greek historians said a man "flourished" at a particular time—more or less the time when he was at the height of his powers, but if a significant event occurred at *any* point of his life this became, naturally enough, the date at which he flourished.

the materialist school of philosophy, which sought to find the ultimate constitution of the world by determining the matter out of which it was made. And in searching for the basic construction of the universe Thales, and hence his followers, accepted the fact that the universe was not only made out of something simple, but that its complexity resulted from changes whereby the basic ingredient of matter (water for Thales) was turned into the diversity of matter we see around us.

The early Greek natural philosophers (the so-called pre-Socratics) were all much concerned with "coming into being" and "passing away" as they put it, meaning thereby change, generation, corruption, life and death, and motion. The fact that they did not distinguish between different kinds of change was later to lead them into profound philosophic problems and some scientific inconvenience, but their emphasis on the problem in itself was most important. Thales—and even more his successors—tried to explain the relation between the earth on which we live, the heavenly bodies, and universal space, thereby introducing for the first time an astronomical element into cosmology, previously magical and religious. Thales, it is true, thought that the earth was a flat disk floating on water like a piece of wood—or perhaps like a plant growing out of the water—but the fact that he speculated upon such matters at all is of immense significance.

Thales chose water as the universal principle and component of things perhaps because water is necessary for life, which is hot though water is cold; perhaps because water can be both liquid and solid. But he apparently never speculated on either how or why water changed into other substances. It was his successor, traditionally his pupil, Anaximander (fifteen or twenty years his junior), who first attempted to solve this problem. Anaximander considered that the underlying stuff of the cosmos was something he called "the boundless," a universal, eternal, unchanging, unlimited, imperceptible, not quite material substance, out of which all matter derives by a selection of attributes or properties. From "the boundless" at the beginning of things opposites were separated by the agency of motion. First came hot and cold, which broke off forming a ring: on the outside hot (i.e., fire); within cold (air); within that again, the earth. The earth began as wet; it dried under the action of the hot, leaving four rings: hot (fire), cold (air), wet (water), dry (earth), the qualities and substances accepted for the next two thousand years as essential in nature. Further, Anaximander produced a theory of the origin of the celestial bodies: he thought the ring of fire broke into rings of the sun, moon, and stars. The sun's ring was twenty-seven times the size of the earth, and the moon's ring nineteen times. These figures are unimportant in themselves, but remarkable as showing that Anaximander regarded it as possible and profitable to speculate upon the physical nature of the heavenly bodies, which clearly for him were not the chariots of the gods, but material bodies susceptible of measurement.

This tendency went further with Anaximenes (about forty years younger than Thales), whose cosmos is increasingly the world around us, and who definitely recognized the difference between stars and planets. He rejected the boundless because it lacked specific properties, and insisted that air or vapor was the underlying constituent of the cosmos. Air rarefied becomes hot (so he thought) and hence turns to fire; air condensed becomes cold and hence turns successively to wind, cloud, water, earth, and stone. Since air is always in motion, change is an ever-present possibility. Air is also breath, and so life. The meteorological emphasis of Anaximenes reflects a growing interest in natural phenomena as well as a realization that the atmosphere is composed of a material substance. And Anaximenes clearly defined the materialist approach which insisted that the cosmos was always to be explained in terms of the matter out of which it was made, a process taking place in space and time.

Thales, Anaximander, and Anaximenes were all Ionians, living on the coast of Asia Minor. Some fifteen or twenty years after Anaximenes, say after 530 B.C., there was developed a rival tradition in the Greek Far West—the colonies of South Italy and Sicily. These philosophers were rationalists, not materialists, and were less interested in the material composition of the cosmos than in its essential characteristics and in such abstruse problems as the nature of change and of existence. Most famous and most influential was Pythagoras (about 530 B.C.) who founded a semireligious, semiphilosophical sect which was to have great influence on the development of Greek thought. As long as the brotherhood lasted (until the mid fifth century B.C.) all Pythagorean doctrine was secret, and hence many of the details of Pythagoras' own thought are obscure. But we do know that he regarded number as the key to the universe and therefore laid special stress upon mathematics. Consideration of his mathematical theories belongs elsewhere (below, pp. 34-35), but certain aspects of them relate to his cosmology. When Pythagoras said that the cosmos was composed of *number,* he invoked a complex series of related ideas. In the first place, his investigation of number theory showed him that all whole numbers could be built up from unity; so he equated the cosmos with unity. Further he taught that one was a point, two a line, three a triangle, and four a solid pyramid, an argument suggested by the custom of representing numbers as points or dots in the sand; this again appeared to show that solid bodies could be "built up" out of number. Pythagoras was much concerned with the subject of proportion, the varying ratios between various numbers; these he related to the lengths of strings which produced musical notes. Certain numbers were thought to have harmonic ratios, again in relation to music, and thus the harmony necessary to the Greek cosmos was provided. Pythagoras did not mean merely that the world was governed by mathematical law, though this is how his doctrine was later interpreted; for him it was the numbers, not the material objects of the Ionians, which were the real world. In the same way, when Pythagoras said that the cosmos was spherical in shape he was expressing a mathematicometaphysical principle, which later philosophers could and did convert into the doctrine that the physical universe was spherical.

Other philosophers of the West took different rationalist positions, but all effectively denied the validity of the material world, believing that too much emphasis had been placed upon the substance of which things were made, not enough on how they came to be as they are. Thus Heracleitos about 500 B.C. insisted that everything is constantly in process of change, that change is the only reality, that one cannot study the material world because it is not the same today as it was yesterday or will be tomorrow. By contrast Parmenides about 475 B.C. insisted that change and motion are illusory. For, he argued, the mind can conceive "being" but not "non-being," so only "being" has reality. "Being" is obviously eternal, unchanging, and motionless, for if it changed it would cease to be "being." This tended, like the doctrine of Pythagoras, to sharpen the distinction between the metaphysical world and the physical world, between the material world of substance and change and the nonmaterial world perceptible, not through the senses, but through the mind.

The criticism of the philosophic rationalists had the effect of rendering materialist concepts more precise, more sophisticated, and more determined. After a lapse of nearly a century there appeared several radically novel and interesting attempts at materialist cosmologies which reflect the rationalists' criticisms and at the same time betray the fact that many thinkers, ignoring cosmological problems, had been quietly investigating physical nature. From about 450 B.C. we find for the first time cosmologies which are clearly based upon some knowledge of the number and nature of the heavenly bodies, and a considerable awareness of the animate and inanimate world. Of these new materialist cosmologies the first and most directly influential is that of Empedocles; the most interesting to the modern is that of Democritos, enunciated about thirty years later.

Empedocles lived in Sicily; he was therefore in a position to be thoroughly aware of the rationalist philosophies of both Pythagoras and Parmenides. Living in one of the richest parts of the Greek world, he took an active role in the political affairs of his native city of Agrigentum, favoring democracy above oligarchy or monarchy. He believed that knowledge was useful as well as intellectually satisfying, and constantly stressed the importance of understanding nature. For some reason, perhaps because he enjoyed the literary exercise, he presented his ideas in a long poem *On Nature,* of which (as with the writings of all the pre-Socratics) only fragments remain. Because of the nature of the exposition, Empedocles' explanations are often couched in poetic language, but his meaning is precisely rational and nonpoetic. In the beginning, there was a spherical universe filled with the four "roots of things," fire, air, earth, and water, which have existed always, but out of which all created things arise. Together with these four "roots," later called elements, there existed the two forces *love* which binds things together and *hate* which separates them. Love is slightly the stronger, but hate is necessary to cause motion and change. The opposition of love and hate separated out, successively, air, fire, earth, and water, and in turn produced night and day, the heavenly bodies, and the universe as we know it. Our world arose by the chance action of the forces of love and hate, and other worlds are conceivable. Empedocles introduced a large number of astronomical considerations: the Moon's light comes from the Sun; the Sun and Moon revolve around the Earth, each turned by an enclosing sphere; eclipses are caused when the lens-shaped Moon passes between the Sun and the Earth; the vault of the heavens is a crystalline sphere (though the universe is egg-shaped), whose motion keeps the Earth at rest in the center. Empedocles' statements show an awareness of the relevance of the heavenly bodies to philosophical considerations that is astonishingly different from anything posed by his predecessors, and the same is true of his biological knowledge. Though he did not practice the mathematical techniques of contemporary Babylonian astronomers, he knew some things that they did not, and his interest in understanding nature, as well as his ability to do so, far transcends theirs; it is doubtful if they would have understood what he was trying to do.

Like Empedocles, Democritos was concerned with solving physical problems by identifying the stuff out of which material objects arise. He was less skilled in astronomy than Empedocles (though tradition has it that he traveled widely in the Near East—he came from Thrace on the northern Aegean shore) but he was an able mathematician and biologist. Democritos was the first important atomist,* and sought in an indefinite number of atoms rather than in a finite number of elements the material basis of the cosmos. For Democritos there were two realities: atoms and the void. Atomism necessarily demands the existence of a vacuum—that very nonbeing which Parmenides had found unimaginable. The Democritean atoms are of many sizes and shapes—perhaps all sizes and shapes—and each substance has a characteristic atom. Thus all atoms of fire are spherical, and since they are literally atoms (uncuttable) they are eternally both fire and spherical. All atoms are endowed with motion, a random and eternal motion similar to that of particles of dust in a sunbeam. This motion brings atoms into contact, when they join together. Though the motion of atoms is random, like seeks like, so some combinations are preferred to others. This is a materialistic philosophy in all senses: there is no guiding force, principle or divinity, and even the mind and soul of man is material, so that sensations and dreams result from the physical impact of atoms. On the whole, the atomic theory of Democritos, though known, was not popular among natural philosophers in the next two or three centuries; it was destined to exercise a fascination through its nonscientific appeal in antiquity, and was to wait for true admiration until modern times.†

The initial inspiration provided by Thales had lasted for

* Democritos claimed Leucippos as his teacher, but the contributions of Leucippos are so vague that even in antiquity no one knew precisely what they were.

† It is hardly necessary to point out that the atomic theory of Democritos has little in common with modern atomism except the name, though the two do have a remote historical connection.

over a century and a half. For all this long stretch of time thinkers had concerned themselves with the problem of the nature of the universe and the question of its ultimate composition. The answers given had been various (they were more numerous than has been detailed above), each satisfactory in some measure and unsatisfactory in another. It would not be surprising if one were to find that the inspiration had run out, and philosophers preferred to concern themselves with other problems. This in fact happened at the very end of the fifth century. The result was to shift the emphasis of philosophic study and effectively to divide philosophers into those primarily concerned with problems of physical nature and those primarily concerned with problems of human conduct. Before this happened, however, the city of Athens was to be the setting for the inauguration of two different and in some senses rival schools of philosophy which have never entirely lost their hold on men's minds. These schools were centered at the Academy and the Lyceum; their inaugurators were Plato and Aristotle.

Athens, the artistic, literary, and political jewel of the Greek world of the fifth century, was hardly the center of philosophic or scientific thought. The great philosophers of the sixth and fifth centuries were not Athenians, and they did not live or teach in Athens. The growth of a democratic form of government rather oddly drew to Athens a good number of teachers of mathematics in the mid fifth century, many of them former members of the Pythagorean brotherhood, now disbanded. They were called "Sophists," since they claimed to teach knowledge; the term came to have its modern dubious connotation because they taught their pupils how to argue so as to win a debate. This was a valuable piece of knowledge to a citizen of a truly democratic state, who might need to argue persuasively in legislature or court of law. The Sophists used the logical arguments of mathematics—geometry already had acquired most of that logical rigor which has given it such a reputation as training for the mind—and were more concerned with scoring a point than with arriving at the truth, to the scandal of many citizens.

One of the Athenians who was most shocked by the Sophists was Socrates, who devoted his life to the pursuit of truth. He was interested in the problems inherent in argument, but more in trying to discover how to find whether an argument's conclusion was true than whether it was convincing. (As a by-product, he helped to develop the art of logical reasoning outside mathematics.) Socrates was not interested in the physical world, but in the world of human society. His aim was to determine how to make men good, wise, and just; to this end he tried to establish the nature of such abstract concepts as goodness, wisdom, and justice. His method was to encourage logical argument by drawing out his pupils (or his opponents) with skillfully devised questions. His pupils—bright young men of the best families in Athens—found his methods fascinating, and they gladly spent many hours talking and listening in his company. What it was like is favorably represented by his most distinguished pupil, Plato, in his *Dialogues*. Plato was profoundly impressed by the ethical content of the philosophy developed by Socrates, by his search for abstract concepts, but above all by his trial and execution by the Athenian citizens in 399 B.C. Socrates was the scapegoat for a city suffering from the aftermath of a demoralizing defeat in war and a disorganized political society; his martyrdom, as he hoped, was ultimately to be an inspiring plea for freedom of speech and thought, in its record by Plato.

The ideas of Socrates have little to do with science: indeed he was opposed to the study of nature because it did not help to an understanding of ethical concepts and what he took to be ultimate reality. In the hands of Plato, however, these ideas, combined with ideas derived from both Parmenides and the Pythagorean tradition, produced a consistent and coherent philosophy and cosmology of great power and great importance for science. Plato (d. 347 B.C.), a young man when Socrates was executed, never entirely recovered from the shock of his master's trial and death. Yet after some years of travel, including practical political experience as adviser to one of the Sicilian kings, he returned to Athens and founded a school, the Academy. Here he lectured and discussed his ideas with numerous pupils, some of whom remained with him for many years. He insisted on much study of mathematics, and some members of the Academy were primarily mathematicians. But his main interest was still the traditional one of the nature of the cosmos.

Plato divided the cosmos into two separate and distinct regions: the world of being and the world of becoming. The world of being he regarded as perfect, eternal, unchanging, the abode of what he called "Ideas" or "forms." All those concepts sought by Socrates—justice, virtue, and so on—existed in perfection in the world of Ideas. So too did there exist Ideas which were the perfection of material objects. The world of becoming—the imperfect, changing physical world—was composed of shadowy, poor copies of the perfect forms of the world of Ideas. To Plato, the world of Ideas was the *real* world; the material world, though seeming real to our senses, was only an illusion. For the senses are unreliable, and play many tricks; Plato had faith only in the intellect, which can be trained to lead us to truth. It is not easy, even for the philosopher, to learn to inhabit mentally the world of Ideas. To put it another way, abstract thought is difficult. The best preparation, Plato thought, was training in mathematics: the geometer, proving theorems about triangles and circles, considers not the ill-defined drawing, but the perfect, abstract, ideal triangle or circle which truly exists in the world of Ideas.

Once again it would seem as if this was a far cry from science, yet there are aspects of Plato's cosmology which are of the utmost importance for science. Plato wrote one book specifically on cosmology: *Timaeus,* a somewhat mystical account, strongly imbued with Pythagoreanism, which was to be the only one of his works known to the early Middle Ages. Here Plato tried to resolve the dichotomy between the world of being and the world of becoming, finding the link in the world-soul. This had been placed by the creator (the demiurge) in the midst of a spherical universe. The world-soul, being self-moved, then created the eternal cosmos and endowed it with perfect motion—circular motion, perfect because it has no beginning and no end and because every point on the circumference is equidistant from the center. At the circumference of the cosmos is perfection; at its center the imperfect earth. From matter (originated by the world-soul) combined with Ideas came substance. Plato recognized the four elements of Empedocles but thought of these as composed of regular solids: fire a pyramid, earth a cube, air an octahedron, and water an icosahedron; these are all resolvable into triangles, which can be formed of lines, which are derived from numbers, which are similar to Ideas. Hence elements, though transmutable, partake of the eternal nature of Ideas. But hence too the key to the understanding of nature is mathematics, now not merely a method of reasoning. Mathematics is the correct technique to use in the search for physical law; the mathematical laws of the universe reflect an underlying cosmic harmony. Plato was also the first to suggest the desirability of finding a geometrical model for the physical universe, a method which was to prove most fruitful in the development of mathematical astronomy. Though Plato's primary aim was not the understanding of physical nature, very many scientists in antiquity and in the modern period have found Platonism a potent inspiration in a search for the mathematical laws and harmonies which govern, or appear to govern, the physical universe.

Plato's successes and failures were to be mirrored in the work of his greatest pupil, Aristotle, who thoroughly understood Plato's doctrines, rejected them for cogent reasons, and formulated replacements of his own. Aristotle was possibly the most astonishing intellectual product of Athenian society. He came from the northern part of Greece, from Macedon, an outlying region where the old form of monarchy had been preserved, whose King Alexander was to become the ruler of the whole Greek world.

Aristotle was born in 384 B.C.; when he was about eighteen, he went to Athens and became a member of Plato's Academy, where he probably remained until Plato's death in 347. Here, obviously, he received a thorough training in mathematics and in Plato's philosophy, of which, however, he became increasingly critical. At Plato's death Aristotle left Athens, lived for a time on the coast of Asia Minor, where he learned a great deal about marine biology, and then spent a few years as tutor to the prince who later became known as Alexander the Great. Aristotle returned to Athens in 335 and founded the Lyceum, which was to prove as famous as the Academy. Here he is said to have lectured while walking about, and the doctrine of Aristotle has been known as "Peripatetic" ever since. Here Aristotle presumably compiled the wealth of material which has come down to us for the use of his pupils. His "published" works—dialogues in the Platonic form—are mostly lost; what we have are his lecture notes and manuals for instruction. So wide is their range, so stimulating and original their content, that anyone who has ever tried to assimilate Aristotle's work as a whole has found it the task of a lifetime.

Though Aristotle wrote on ethics, politics, and literary criticism, the bulk of his work dealt with various aspects of natural philosophy, broadly interpreted. At one end of the spectrum he treated scientific method; at the other he transcended natural philosophy in his works on metaphysics. In between he formulated a scientific cosmology which was to provide man's picture of the universe for nearly two thousand years, at least intermittently, a cosmology bound together by a coherent underlying physics and theory of matter which in scientific scope and imagination goes far beyond the work of his predecessors. He also composed the first extant works dealing with theoretical biology, and his zoological writings are astonishing in their range and accuracy.

Heir as he was to a long tradition, Aristotle endeavored to complete half-begun problems and to settle major issues. Like every good teacher he was at immense pains to explain his chosen position: he not only gave reasons for his decision, but carefully refuted any actual or potential opponents. He even felt the need to examine and discuss the underlying logical construction of his thought. So, half a century after Socrates had confused the citizens of Athens (and perhaps himself) in a way only possible when an understanding of the precise powers of a logical argument is lacking, Aristotle wrote a major work on logic, and thereby founded a new branch of philosophy. Partly because formal logic had, previously, been chiefly associated with mathematics, Aristotelian logic is strictly nonmathematical. He was more interested in the logic of words, and developed immensely important methods of distinction of semantic types, as well as describing and analyzing the now familiar syllogistic form of argument. Just as he classified syllogisms, and analyzed the difference between different types, so he tried to analyze and classify all subjects, from political constitutions to forms of motion, as well as the animal kingdom. Classification was not an invention of Aristotle's, but his taxonomic interests were complex and far superior to the simple dichotomies beloved of Plato.

Aristotle's cosmological construction of the universe is so much more precise and sophisticated than Plato's as to seem almost divorced from the preceding two centuries' speculations. Yet though the Aristotelian cosmos is recognizably more scientific than those of the pre-Socratics, there is still a marked kinship; it is just that Aristotle filled in the gaps and solved the problems posed by those before him until the smoothness of the finished product almost conceals the details of manufacture.

The Aristotelian universe is a cleverly contrived mechanism, though Aristotle himself would not have been capable of recognizing its mechanistic character. It consisted of a series of concentric, nesting spheres, the outermost one, the *primum mobile,* at rest, the inner spheres moved by the *primum mobile* except those near the center, which were at rest again. These spheres were solid, concrete objects, the crystalline spheres of the Middle Ages, composed of a perfect material which was pure and unchanging. Indeed, the whole celestial region was perfect and immutable—the world of being, eternal and unchanging. (Aristotle rightly pointed out that it was a logical contradiction to imagine, as Plato had done, that the world could be created and eternal; it must either be created and doomed to destruction, or have existed always and be eternal. He preferred the latter alternative.) The world of "becoming" Aristotle confined to the terrestrial or sublunary region, the sphere of the moon marking its boundary. Here, at the center of the universe, was the changing, finite, terrestrial world as we know it, marked by generation and corruption and transmutation. And just as the celestial region was composed of the incorruptible quintessence, the terrestrial world was composed of the four elements, earth, water, air, and fire. At the very center, as befitted its heavy and inert nature, lay the sphere of earth: and as the Earth, the terrestrial globe on which we live, is mainly composed of the element earth, it naturally remains at the center of the universe. Beyond the sphere of earth was to be found the sphere of water; beyond that again the sphere of air; and beyond these, that of fire, the last of the sublunary elements. In the Aristotelian universe the sun, stars, and planets are not composed of fire, but of the fifth essence; they shine because the motion of their spheres produces friction of the air far below, and friction of the air produces light and heat.

Each of the four elements had associated with it certain qualities. Aristotle, rejecting Plato's doctrine of Ideas, retained Plato's distinction between *form* and *matter*. But whereas Plato held that pure forms exist and are to be found in the world of Ideas, Aristotle declared that pure forms exist only in the mind as a product of abstract thought. Forms exist in the material world only in association with matter. The elements themselves possess certain forms or qualities: earth is dry and cold, water wet and cold, air wet and hot, fire dry and hot. So the formal cause of any object is the combination of properties which make it what it is. And as every substance in the universe contains some of each of the four elements, every substance contains the four basic (but contradictory) forms, and so can be changed into some other substance by a transmutation of forms. This theory, much developed, was to serve the alchemists in good stead.

Since each element has a natural place in the universe, a sphere to which it belongs, so, reasonably enough, each element must seek to return to its natural place if removed from it. Thus water raised into the air falls as rain, while water which is below the surface of the earth rises in the form of springs. Water is thus both heavy and light; for to be heavy means to have a tendency to fall, light to have a tendency to rise. Fire is always light, earth always heavy, while air, like water, can be either. All elements have weight in their own spheres; in fact Aristotle discussed methods for weighing air. Obviously, the gravity of a stone—its tendency to fall to the earth—is only one example of natural motion. This motion, Aristotle believed, always took place in a straight line, the shortest distance between two points. (The celestial regions had a separate physics; their natural motion was circular.) Not unnaturally, Aristotle assumed that the more there was of an element, the faster it moved to get to its natural place. In other words he assumed that heavy bodies fall faster than light bodies. It was not that he did not recognize the presence of air resistance, for he did, quite explicitly. The thinner the medium, Aristotle perceived, the faster the motion, so that, for example, a stone falls faster in air than in water. In a very thin medium the stone would fall very fast indeed; in no medium—i.e., a vacuum—the stone would fall at an infinite velocity, or instantaneously. Which is absurd, as Aristotle rightly says. Hence, for this as for other reasons, a vacuum cannot possibly exist.

Just as a medium was necessary for the existence of natural motion, so it was, Aristotle thought, for forced or projectile motion. For why otherwise should a stone continue to move after it leaves the hand or sling that propels it? He argued that perturbations must be set up in the air—like the ripples visibly set up in water—and these must assume the role of propulsion, so that the projectile is continuously pushed, though with weakening force. (As air is so thin, and manifestly apt for motion—as the existence of winds and air currents shows—Aristotle supposed that the air could continue to cause motion for a relatively long time.) Finally, the air will no longer be able to maintain its thrust, and the projectile will fall straight to the earth under the force of natural motion.

It is clear from our vantage point that these are not very good theories of moving bodies, for bodies do not fall at speeds proportional to their weights (though manifestly very heavy bodies do fall through air faster than very light ones). Nor do we accept Aristotle's view of what keeps a body moving. In fact, where Aristotle insisted that motion could take place only through a resisting medium, we regard the medium as extraneous. Aristotle would have regarded our view as far too Platonist, far too nearly a discussion of motion as it might occur in the perfect world of ideas. And for the same reason Aristotle minimized the role of mathematics in physical science, because mathematics treats of pure form, whereas in the physical world form is always combined with matter. Aristotle saw—and was perhaps the first to do so—that a theory of motion was needed for the terrestrial as well as the celestial regions; and he set about framing such a theory with logical thought and a clear reference to experience. For in rejecting the mathematization of the world, together with Plato's doctrine of ideas, Aristotle was endeavoring to restore sense experience to the important place in physical speculation denied it by the philosophic rationalists and after them by Plato. This endeavor is more apparent in Aristotle's biological writings than in his physical works, but it is omnipresent nevertheless. The unquestioning adherence to the word of "the master of those that know" insisted upon by medieval schoolmen should not blind us to the fact that Aristotle had never intended his system to be final truth. The true Aristotelians are the later generations of Greek scholars who carried on in those areas where he had begun, not afraid to criticize as well as to extend. The Aristotelian system is a marvelously coherent whole, such as was not to be achieved again until Descartes in the seventeenth century attempted a similar, though less extensive, system. Though its great power had a somewhat stultifying effect on those who came upon it unprepared, it could be, and in the century after Aristotle's death truly was, an inspiration and a model which helped lead to the development of the sophisticated science of the Hellenistic Age.

Chapter 3

THE MATHEMATICAL WAY IN ANTIQUITY

In many respects, Greek physical science was as mathematical as modern science is. The Pythagorean tradition that number was the ultimate reality led men to feel that the universe was governed by mathematical law. Though it was not until the later years of Plato's Academy that this concept was formalized, it had long existed as an undercurrent and was readily made explicit in the great age of Greek science coincident with the flowering of the Hellenistic culture developed in the wake of Alexander's conquests.* But pure mathematics is far older; Thales is credited with knowledge of geometry, and the basis of Pythagorean doctrine was number theory. For the former, Thales may (as tradition claimed) have drawn on Egyptian technique, though he must have transmuted it by the action of his mind; the latter was original with the Greeks.

Nothing shows more vividly the difference between the "mathematics" of Babylonia or Egypt and that of Greece

* Alexander died in 323 B.C., having temporarily established Greek rule throughout Asia as far east as India. When he died his empire disintegrated into four main fragments: the Greek mainland; Asia Minor; the Seleucid kingdom, which controlled Persia, Mesopotamia, and the Mediterranean coast; and Egypt. In all these areas the cities were Hellenized and Greek became the common tongue. The most important intellectual center was the new Greek city of Alexandria at the mouth of the Nile.

than what each culture chose to do with numbers. The Babylonian and Egyptian mathematicians were calculators and manipulators, devising number systems suitable for computation and problem solving. The Greeks were content with a far more primitive alphabetic number system, cumbersome for calculation, though in fact they performed multiplication and division as we do. For everyday affairs, however, the Greeks used an abacus, which was probably a Greek invention. The abacus uses the principle of positional, decimal numeration; only the result needs to be written down, and the kinds of numerals used have nothing to do with the mechanics of calculation. Anyone could learn to calculate with an abacus; so reckoning (*logistic*) was no longer a learned art as it had been to Egyptian and Babylonian scribes.

What interested Greek thinkers were the abstract properties of numbers, and their *arithmetic* (number theory) was a highly developed branch of mathematics. They discovered the properties of odd and even; square, cube, and triangular numbers (reflecting their representation by dots or pebbles); and perfect numbers (those, like 6, which are the sum of their factors). The Pythagoreans were keenly interested in the theory of proportion, that is in the ratios between various numbers, and were in time led to the discovery of irrational numbers. This was said to have had a devastating effect upon the Pythagorean brotherhood, only mitigated when the theory of proportion was further developed to include irrationals. It was their interest in proportion and harmony which led the Pythagoreans to regard music as a mathematical art (a concept to appear again in the quadrivium of the medieval university), and their "music of the spheres" was initially represented by mathematical statements about ratio and proportion of sizes and distances.

Except for some (but not all) parts of number theory, Greek mathematics was entirely geometrical. Here again the Greeks were primarily interested in abstract concepts about ideal figures of circles, triangles and parallelograms, and very little interested in solving problems of areas. Not that they were unable to solve the kinds of problems beloved by the Egyptians and the Babylonians; by the end of the fifth century B.C. they could solve linear and quadratic equations geometrically, though they were always interested in the method rather than in the result. Their methods became so generalized that it took little skill to solve a problem once the method was known, and simple manuals of how to do so were not regarded as worthy of preservation by later compilers.

The difference between Greek geometry and the work of Egyptian mathematicians is excellently demonstrated by the traditional stories of Thales. He is said, for example, to have stated that a circle is bisected by its diameter, and that the base angles of an isosceles triangle are equal. These are facts of which Egyptian computers often made use; it appears that what Thales began to do was to make general statements about circles and triangles. The same thing is observable in connection with the traditions of Pythagoras and the early Pythagoreans. Both Egyptian and Babylonian calculators made use of the peculiar properties of a right triangle with sides 3, 4, 5. The Pythagorean theorem (that the square on the hypotenuse of a right triangle is equal to the sum of the squares on the other two sides) is a generalization, applied to all right triangles (and incidentally indicates a complete recognition and classification of triangles). Further, Pythagoras is said to have *proved* this theorem. And indeed it was not only generality that made Greek geometry novel; it was the development of the concept of proof. Assuming a relatively few common notions, Greek geometers began to build up an established body of proved theorems. In doing so, they invented various forms of mathematical argument (including the familiar *reductio ad absurdum*) and also compiled a growing cumulative body of mathematical knowledge. By about 450 B.C. the earliest *Elements* (i.e., textbook) of geometry was written, the first of a series. This series ended when, about 300 B.C., Euclid wrote the version which so successfully replaced its predecessors that they were lost while Euclid's *Elements,* virtually unchanged, remained a standard introductory textbook until less than a century ago.* Similarly, geometry retained the status of perfect training in abstract thought given it by the Pythagoreans and adopted by Plato.

Elements, of course, were concerned only with elementary geometry, and mathematicians had gone well beyond the rudiments before these were systematized. One very popular problem—which was to lead to several most in-

* Elementary geometry in medieval universities, as in modern schools, included the first six books of Euclid's *Elements,* including the general theory of proportions. Books VII and IX deal with number theory, VIII with geometrical series, X with irrationals, and XI–XIII with solid geometry. There is nothing on conics, which was too advanced for inclusion here.

teresting developments—was the attempt to "square the circle," that is, to find an area bounded by straight lines equal to the area of a given circle. (This undertaking was, in fact, an impossibility, though the Greeks were not to know that.) It led to researches into the quadrature (or squaring) of various curvilinear plane figures, especially by Archimedes (d. 212 B.C.). It also led to the "method of exhaustion," an attempt to determine the area of curvilinear figures by inscribing within them rectilinear figures of calculable magnitude. The method was to draw, say, a hexagon; then draw a triangle on each of its sides, and so on. A similar method of dividing up irregular areas into small segments was to be used by Archimedes to determine the area under a curve; eventually, in the seventeenth century, this method led to the infinitesimal calculus. But first the question of infinite divisibility had to be solved: was it possible to continue to divide up an area until nothing remained over? That was a question hotly debated—by Democritos, among others—and dramatized by Zeno's paradoxes. Of these the most famous is that of Achilles and the tortoise, one of a pair which show that *if* magnitudes are infinitely divisible, then motion is impossible; whereas other paradoxes showed that if space and time are composed of indivisible parts, then motion is also impossible. The answer of Plato's pupil Eudoxos is that the mathematician need not attempt to reach the infinitely small; he can be content to divide a magnitude until he reaches one "as small as we please." So although Greek mathematics did not arrive at a clear definition of a limit, it approached very close to it.

One other aspect of advanced geometry must be mentioned: the study of curvilinear figures more complex than the circle. The most important of these are conic sections, first studied before the time of Euclid, whose own treatise has not survived. Archimedes worked on the parabola, and discussed the ellipse. The basic treatise, however, was the *Conics* of Apollonios of Perga (third century B.C., about twenty-five years younger than Archimedes). This is a complete, thorough, and systematic investigation of the properties of conics, so advanced that it was still the best work on the subject in Newton's day.

Various aspects of the mathematics of the third century B.C. continued to occupy mathematicians for the next six centuries. No new subjects were introduced, however, except for the attempt in the first centuries of the Christian era to introduce arithmetical methods for the solution of certain problems hitherto solved geometrically.* This primitive algebra may well have been influenced by late Babylonian developments (c. 300 B.C.). The great synthetic text was the *Arithmetica* of Diophantos (second or third century A.D.), which is notable for the first use of a symbol for the unknown, as well as for its understanding of complex equations.

In spite of the Greek insistence on abstract concepts in mathematics, applied mathematics had never been rejected. The Pythagoreans, indeed, had ranked astronomy with the mathematical sciences of arithmetic, geometry, and music, holding it to be a form of geometry since it dealt with magnitudes, but differing from geometry in that it dealt with magnitudes in motion. Presumably the Pythagorean declaration that the Earth, like the universe, is a sphere stemmed from considerations of mystical mathematics. The early Pythagoreans had a physical model of the universe, recognizing that the motions of the planets, Moon, and Sun were periodic; they made no attempt at calculation of this periodicity, and their model was removed from reality by their belief that the center of the universe was a "hearth," and that planets, stars, the Earth, and a symmetrical counterearth all moved around this central fire.

The first serious enunciation of the possibility of a mathematical model of the universe was Plato's. By the early fourth century B.C. enough was known about the motions of the planets—whether from direct observation or through borrowings from Mesopotamia—to make this possible. Plato's notion derived from his devout belief that the planets must move in perfect circles—for the Platonic heavens, like the Aristotelian heavens, were closer to perfection than were the earthly regions—coupled with the recognition that whereas Sun, Moon, planets, and stars rise and set regularly every twenty-four hours, the motions of the planets against the backdrop of the stars were exceedingly irregular.† This is particularly true of the outer planets (Mars, Jupiter, and Saturn); because they move more slowly about their orbits than the Earth does about its, the Earth will overtake and pass them at regular intervals, making their paths appear to loop back upon themselves. This is the phenomenon known as *retrograde motion* and it was, in Plato's day, the phenomenon which most needed explaining. Plato became convinced

*True, trigonometry dates from the second century B.C., but as a branch of astronomy.

† Throughout Greek antiquity the planets were seven: Saturn, Jupiter, Mars, Venus, Mercury, the Sun, and the Moon.

that one could account for all irregularities—"save the appearances" as it was commonly put—in terms of perfect circular motion, a difficult but not impossible problem in the then state of mathematics. And so useful and satisfactory a concept was this that it was to remain the driving force in mathematical astronomy until the time of Kepler.

The first solution to this problem, mathematically very ingenious, was devised in the Academy by the mathematician Eudoxos, whose life spanned the first half of the fourth century B.C. Eudoxos assumed that the observed motion of the planets could be represented mathematically if each planet were imagined to be situated on the equator of a sphere rotating with uniform speed. He further assumed that the poles of this planetary sphere were not stationary, but were carried by a larger sphere, in turn rotating about its own axis with uniform speed. All spheres were assumed to have the same center, but the poles of each sphere were different and so the rotation of each sphere was different. By assuming three spheres for the Sun and Moon, four for each of the planets (one sphere in each case accounting for the daily rising and setting) and one for the fixed stars, Eudoxos was able to represent all the known motions of the heavenly bodies, even retrograde motions. It must be remembered that the Greeks used geometry where we use algebra; the spheres of Eudoxos were mathematical spheres, not material ones, and represented a mathematical description of the celestial motions, not a physical model. Eudoxos' system fulfilled the requirements of Plato and had the further advantage of emphasizing the central position of the Earth.

Not surprisingly, in view of his belief that there was a profound difference between the mathematical world of pure form and the physical world of form and matter intermingled, Aristotle was disturbed at the way in which Eudoxos solved Plato's problem. He saw its mathematical power, and approved of its retention of circular motion, but deprecated its purely mathematical structure. It occurred to him that the concentric spheres of Eudoxos would be equally effective in explanation if they were material spheres instead of mathematical ones. They would, of course, be composed of some perfect and unchanging material, suitable to the celestial regions, absolutely transparent and hence invisible: the quintessence, later denominated *crystalline*. The corporeal spheres rotated because circular motion was natural in the heavens. But their corporeality, together with Aristotle's conviction that the universe must be studied as a whole, led him to conclude that these spheres which provided planetary motions must be physically linked in some way. Connecting all the various inner spheres to the outermost *primum mobile* (first mover) compelled Aristotle to introduce a number of complexities and to increase the total number of spheres, but his "mechanical" system had the notable advantage of producing for the first time a picture of the universe as a whole with interlocking parts, each following natural (though not mathematical) law.

The beauty and conceptual utility of Aristotle's version of the Eudoxan system was not fully appreciated in antiquity; it was not until the Middle Ages that it was elevated into dogma. To the scientists of the Hellenistic period, this was but one way of saving the phenomena. Other methods were suggested. A Pythagorean contemporary of Aristotle, Heraclides of Pontus, noted that if the Earth rotated on its axis, this would account for the daily rising and setting of the planets and the fixed stars (which in that case would be truly fixed); he may also have suggested that the inner planets, Venus and Mercury, which always appear near the Sun, revolve about the Sun rather than about the Earth. The ideas of Heraclides were adopted by various astronomers throughout the subsequent centuries, but we have very little idea of how his contemporaries appraised his suggestion. Even more tantalizing is the theory of Aristarchos (a contemporary of Archimedes): he held that the fixed stars and the Sun are motionless, but that the Earth revolves around the Sun on a circular path. Presumably he also thought that the Earth rotates daily on its axis. Aristarchos wrote a book about his hypotheses, as Archimedes called them, but it is lost; our major source is a brief description by Archimedes, who wanted to compute the size of the largest possible universe; that of Aristarchos had to be larger than that of most astronomers in order to explain the absence of parallax.* This is most astonishing, and one would like to know more, especially since Aristarchos was right. But very probably the hypotheses of Aristarchos were presented as mere hypotheses. Certainly he could have had no more physical proof than Copernicus was to have almost exactly eighteen centuries later. And it is equally probable that he did not work

* If the Earth moves, the stars ought to appear to have different relative positions at different times of the year. Parallax can be observed by anyone who walks across the front of a classroom or theater and notes how the chairs in the various rows appear to shift position with respect to the wall behind them.

out the mathematics of the situation. We should not be surprised therefore to find so few references to the Aristarchan system in antiquity, nor need one take seriously the light remark of Plutarch some three centuries later that Aristarchos was regarded by some as too impious to be taken as a guide. The failure of Aristarchos was the failure of an inspired and ingenious but implausible and unsupported guess in the face of mathematical and physical reasoning which consorted with the evidence of the senses.

Interest in mathematical astronomy led to interest in observational astronomy. Whether the observations were Greek or Babylonian is unimportant, but it remains true that in the third century B.C. Greek astronomers became simultaneously aware of the varying brightness of some of the planets (especially of Mars) and also of certain complex periodic variations in their motions. Neither of these could be represented by the system of concentric spheres, nor were they dealt with by Heraclides or by Aristarchos. New mathematical methods were accordingly devised which proved so successful on a geostatic system that there was no need to try to adapt them to the Aristarchan system. (That was first done by Copernicus.) The first device was the use of the *eccentric:* that is, of a circle or sphere whose center is not the Earth, though the circle or sphere may be said still to "go round" the Earth. The second, even more successful device, associated with the name of Apollonios the mathematician, was the combination of circles known as *epicycle* and *deferent*. The planet was assumed to lie on the circumference of a small circle, the *epicycle,* whose center lay on the circumference of the larger circle, the *deferent*. The center of the deferent might be either the Earth or some other point *C*, in which case the deferent was eccentric. Both the epicycle and the deferent rotated about their centers (Figure 1). It is easy to perceive that the planet, *P*, will follow a looped path, giving the requisite retrograde motion more easily than with the use of spheres. Further, by adjusting both the size of the circles and their velocities to each case, a very satisfactory mathematical representation of the motion of the planets can be achieved. Much later a further refinement was introduced, the *equant point,* which explained why the velocity of a planet is not uniform throughout its path.* For the equant point—the point about which the planet describes equal

* As Kepler's second law indicates, the velocity is faster when the planet is nearer to the Sun.

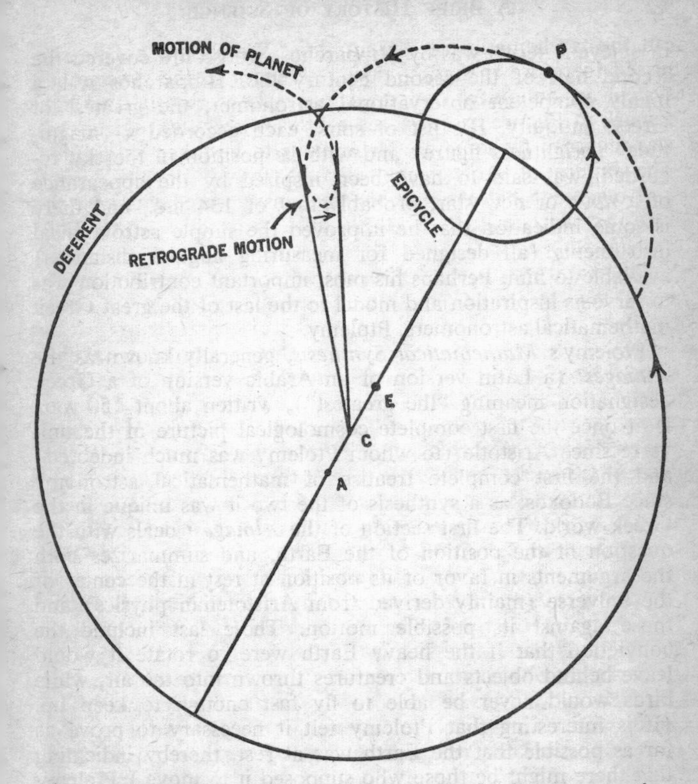

Fig. 1. The mathematical representation of planetary motion. *P* represents the Planet; *C*, the center of the deferent; *E*, the Earth (when the deferent is eccentric); *A*, the equant point.

angles in equal times—was removed from the center of the circular deferent. From a geocentric view this was stretching things, but one could argue that uniform circular motion *was* being retained, and certainly the appearances were thereby more adequately saved.

The epicyclic system was purely mathematical; it said nothing about any mechanical spheres (though it did not in any way exclude their use), nor about any mechanical model. Its mathematical application was difficult and complex. The first complete discussion of the motions of the Sun and Moon in epicyclic terms was by Hipparchos, whose life covered the second half of the second century B.C. Hipparchos was a highly competent observational astronomer, the greatest of Greek antiquity. His list of stars, each accorded a "magnitude" (brightness figure) and with its position in the sky recorded, was said to have been inspired by the appearance of a *nova* or new star, probably that of 134 B.C., and there is some indication that he improved the simple astronomical instruments (all designed for measuring angular distances) available to him. Perhaps his most important contribution was to serve as inspiration and model to the last of the great Greek mathematical astronomers, Ptolemy.

Ptolemy's *Mathematical Synthesis*, generally known as the *Almagest* (a Latin version of an Arabic version of a Greek designation meaning "the greatest"), written about 150 A.D., is at once the first complete cosmological picture of the universe since Aristotle (to whom Ptolemy was much indebted) and the first complete treatise of mathematical astronomy since Eudoxos; as a synthesis of the two it was unique in the Greek world. The first section of the *Almagest* deals with the question of the position of the Earth, and summarizes both the arguments in favor of its position at rest in the center of the universe (mainly derived from Aristotelian physics) and those against its possible motion. These last include the conviction that if the heavy Earth were to rotate it would leave behind objects and creatures thrown into the air, while birds would never be able to fly fast enough to keep up. (It is interesting that Ptolemy felt it necessary to prove as far as possible that the Earth was at rest, thereby indicating that there might be those who supposed it to move.) Ptolemy retained Aristotle's sublunary spheres and the distinction between celestial and terrestrial physics; indeed there is good evidence that he believed in the existence of crystalline spheres for each planet, though in the *Almagest*, which is purely mathematical, he does not say so. The bulk of the *Almagest* is devoted to the mathematics of the planetary motions. Ptolemy's discussion of the motion of the Sun is identical with that of Hipparchos, but he improved on the lunar theory of Hipparchos, and added detailed mathematical accounts of the motions of the planets. Here he introduced the equant point, which was unknown to Hipparchos and may have been invented by Ptolemy. Never before had Plato's injunction to search for mathematical means of representing the observed motion of the heavenly bodies in terms of circular motion been so thoroughly and carefully followed. Ptolemy worked out the mathematics of his system is complete and satisfactory detail. No better tribute has ever been paid to the *Almagest* as a work of mathematical elegance than the fact that Copernicus, attempting to disprove Ptolemaic astronomy, took the *Almagest* as a model and followed it faithfully point by point and section by section in order to demonstrate the power of his new system.

Planetary theory was a most important contribution of mathematics to an understanding of the structure and harmony of the universe. Its fundamental concept—that the apparent irregularities of the motion of celestial bodies could be reduced to mathematical law—was bold and daring. Yet the Greek mathematical astronomers of the Hellenistic age, seeing that it offered so proper and promising an approach to a knowledge of the heavens, readily accepted its challenge. From the time of Eudoxos the aims and methods of mathematical astronomy had been almost universally accepted and were steadily developed. In spite of the intricacy of the problems it invoked, however, it by no means absorbed all the energies of the mathematical astronomers of antiquity. With equal boldness they set out to measure the universe, thereby further reducing it to an intelligible entity subject to mathematical treatment. Earlier discussions of the sizes of the heavenly bodies (by various pre-Socratics) had been mere ingenious guesses. By the third century B.C. sophisticated methods capable under ideal conditions of producing fairly exact answers had been devised, and presented in mathematical form.

One of the most interesting of such attempts is the subject of the only surviving work of Aristarchos, *On the Sizes and Distances of the Sun and Moon.* The presentation

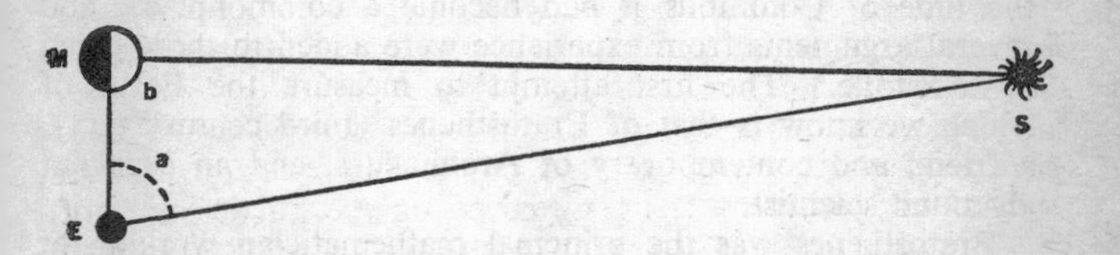

Fig. 2. Aristarchos' measurement of the sizes and distances of the Sun and Moon.
M represents the Moon; *S*, the Sun; *E*, the Earth; *b*, a right angle; and *a*, the angle to be measured. This is quite out of scale; *a* is, in fact, very nearly a right angle.

in this little treatise is strictly mathematical: first come six hypotheses, then follow three propositions, complete with proof, exactly as in Euclid's *Elements*. Aristarchos' "hypotheses" are, however, an interesting combination of astronomical assumption ("that the Moon receives its light from the Sun," for example) and actual astronomical measurement. His method was as follows (Figure 2). Given the hypothesis "That, when the moon appears to us halved, the great circle which divides the dark and the bright portions of the moon is in the direction of our eye,"[1] then the Earth, Moon, and Sun form a right triangle when the Moon is exactly half full. The angle between the Moon and Sun (angle *a*) can readily be measured. Having previously determined the breadth of the Earth's shadow (measured during lunar eclipses) and measuring the angular diameter of the Moon, Aristarchos was able simultaneously to determine the relative sizes and the relative distances of the Sun and the Moon, since the lines connecting them form sides of a triangle, and he knew the sizes of their opposite angles. His conclusions—that the Earth–Sun distance is eighteen to twenty times the Earth–Moon distance, and that the diameter of the Sun is between 6⅓ and 7⅙ times that of the Earth—are not particularly accurate; observational difficulties combined with the fact that a small error in measurement makes a large difference in the result. The reasoning and the geometry are impeccable.

Another fascinating example of this kind of ingenious geometrical astronomy is the attempt to determine the exact size of the Earth's circumference. (It will readily be apparent that there was no question of whether the earth was a sphere; no scientist after the time of Aristotle doubted it. Nor did anyone who read Aristotle in the Middle Ages; by the time of Columbus it had become a commonplace, and several arguments from experience were added to those given by Aristotle.) The first attempt to measure the Earth of which we know is that of Eratosthenes (third century B.C.), a friend and contemporary of Archimedes, and an excellent all-round scientist.

Eratosthenes was the principal mathematician working at the Museum at Alexandria, a research institute supported by the Greek rulers of Egypt and the chief center of science and scholarship in the Hellenistic world. He reasoned as follows.

[1] Superior numbers throughout text refer to sources for quotations. These references are enumerated according to chapter and begin on p. 328.

Alexandria and Syene (two Greek cities in Egypt, the latter near modern Aswan) lie on the same meridian circle (i.e., have the same longitude), and the distance between them is known. If we can measure the angle they subtend at the center of the earth (*ACS* in Figure 3), then we will know

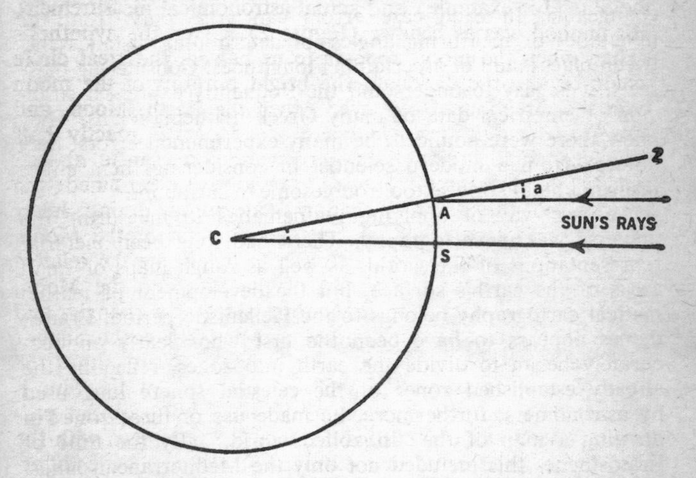

Fig. 3. Eratosthenes' measurement of the circumference of the Earth.
Z represents the Zenith; *C*, the center of the Earth; *A*, the city of Alexandria; and *S*, Syene. The angle *a* is the angle to be measured; it is equal to the angle *ACS*.

what portion of the whole circumference the arc *AS* is, and since the length of *AS* is known, the length of the circumference may readily be calculated. To do this Eratosthenes needed two assumptions, one observation, and a couple of geometrical facts. His assumptions are that the sun's rays falling on different points of the earth's surface are always parallel, and that at Syene at the summer solstice the sun casts no shadow (i.e., the sun is at the zenith). Now, by observing at Alexandria a shadow cast by a sundial, the angle between the Sun and the zenith can be measured. When this has been done, the angle *ACS* can be shown to have the same value, since, as is apparent from the figure, one needs only to know that vertical angles are equal, and that when a straight line cuts two parallel lines the alternate interior angles are equal. Eratosthenes found that *AS* is about

1/50 of the earth's circumference.* Two centuries later a similar computation was undertaken in which the angular measurement involved is that of the height of a star above the horizon at two places on the same meridian. Clearly, both these methods are perfectly sound and geometrically interchangeable. In each case errors can arise only from the difficulties of measuring angles, of determining distances between places, and of ascertaining longitudes. Considering these facts, it is easy to understand the frequently expressed suspicion of empirical data of many Greek mathematicians. They knew there were bound to be many experimental errors; they differed from a modern scientist in considering them either insurmountable or else too troublesome to surmount.

Another way of applying mathematics to measuring the universe was in cartography. There had long been pictorial representations of the world, as well as rough maps of small areas of the earth's surface, but the development of mathematical cartography belongs to the Hellenistic period. Eratosthenes appears to have been the first who clearly and accurately began to divide the earth into zones, reflecting the already established zones of the celestial sphere long used by astronomers; furthermore, he made use of these zones in drawing a map of the "inhabited world." (By the time of Eratosthenes this included not only the Mediterranean lands, but part of northern Europe, including the British Isles, Africa, the Arabian peninsula, India, and, less securely, the lands north of India, and east of the Caspian Sea.) Later, latitude and longitude (based on the breadth and length of the Mediterranean) were introduced, though there were always difficulties in getting anything like an exact determination of either. The most complete Greek treatise on mathematical geography (there are a number on descriptive geography) is Ptolemy's *Geography*. This includes all aspects of mathematical geography, as well as many tables giving the latitudes and longitudes of numerous cities of the Greek world. (From these it has been possible to reconstruct Ptolemy's maps, which have not survived.) Not only did Ptolemy provide maps; he also furnished a long discussion of how to draw maps, with an explanation of the problem of *projection*. This arises because the earth is a sphere; we draw a map on a plane surface although it is not possible to flatten a hemisphere into a plane. Though maps without projection are

* The exact value of the unit of length used by Eratosthenes is unknown, but it is obvious that the notion of the earth's size that he derived is roughly correct.

satisfactory for very small areas, Greek geographers had long been dissatisfied with plane maps in Ptolemy's day. He recommended a number of methods, including the use of a conical projection: here points on the spherical earth are transferred to the base of a cone, which can be flattened into a plane if cut. The *Geography* is an elegant mathematical treatise, destined to exert an enormous influence on European cartography in the Age of Discovery.

Physics, especially mechanics, also flourished in the Hellenistic Age, taking its departure either from the subjects discussed by Aristotle or from a Platonic belief in rigorous mathematical law. The ideas on motion discussed by Aristotle were further developed, and even criticized, in the *Mechanical Problems,* probably by Strato (c. 287 B.C.) who was the third head of the Lyceum and also worked in Alexandria. Strato wrote a treatise *On Motion* in which he discussed the fact that the speed of a falling body increases as it falls, and appears to do this uniformly throughout its fall; in the *Mechanical Problems* he also raised the question of why it is easier to move a body already in motion than one at rest. The *Mechanical Problems* is primarily a treatise on statics and on four of the five simple machines,* handled more qualitatively than was common in later Greek discussions. Starting with a definition of mechanical problems as those whose method belongs to mathematics and whose application belongs to natural philosophy, Strato proceeds to consider the law of the lever (which he knows accurately) as a case of equilibrium and demonstrates its validity by showing how an alteration would produce motion about the arc of a circle. Other simple machines—wedge, pulley, steelyard, balance, as well as the oar, rudder, and dental forceps—are reduced to special cases of the lever. Further, Strato makes use of the parallelogram of forces, or rather, the parallelogram of velocities, which he states quite explicitly. Implied in his discussion of projectile motion is not only doubt about Aristotle's theory, but a suggestion that an object thrown must offer resistance in the direction from which the thrust comes. All these questions—and more—continued to occupy the attention of a number of Greek physicists; they are dealt with in a manner almost identical with that of Strato, though in a more overtly mathematical dress, by Hero of Alexandria (first century A.D.) in his *Mechanics*. This was, however, less influential since it was never translated into Latin, whereas

*Lever, inclined plane, pulley, windlass, and screw.

the *Mechanical Problems* had the advantage of being included as a treatise of Aristotle's along with the genuine works.

The treatises on mechanics by Archimedes are of quite a different nature. So far as we know he had no interest in problems of motion; his interest lay rather in statics. In *On the Equilibrium of Planes* he proved the law of the lever in a series of rigidly geometrical propositions of great mathematical elegance, and also considered a wide range of center-of-gravity problems. Here, as in his hydrostatics, he deals with *magnitudes,* not with bodies, thereby rendering his work purely mathematical, as well as avoiding the Aristotelian objection to mathematical physics. Archimedes never claims to be dealing with real bodies, only with the ideal magnitudes of the geometer's ideal world. Thus, for example, he can discuss the centers of gravity of plane figures (which have no material existence) with complete validity. In his work on hydrostatics, represented by the treatise *On Floating Bodies,* he similarly deals with pure mathematical magnitudes, in this case three-dimensional figures, and again proceeds with great geometrical rigor; here, however, he was compelled to approach the material world, since he considers his figures to have weight—necessarily so since he is interested in the problem of specific gravities. But a large part of the work is devoted to a consideration of the stability of floating segments of a paraboloid, a problem of little direct practical concern. Nothing is more unlike the coldly formal and elegantly geometrical rigor of Archimedes' actual books than the legend attached to the supposedly fortuitous discovery of the principle of Archimedes (that a solid weighed in a fluid will be lighter than when weighed in air by the weight of fluid displaced), a story popular since antiquity.

After Archimedes, Greek mechanics developed in the direction of practical application. From simple machines Greek mechanicians went on to describe fairly complicated mechanical contrivances: the increasingly complex war machines of late Hellenistic and early Roman times, automata and pneumatic and hydraulic devices in particular. Both Philo of Byzantium (second century B.C.) and Hero of Alexandria tried to consider the theoretical basis of the mechanical contrivances they described, reducing them where possible to simple machines. These in turn Hero reduced to the principle that the smaller the force which lifts a given weight the longer the time it will take. Except for war machines, water clocks, and (in the case of Hero) pumps, most of these contrivances were applied to toys and tricks. Hero, for example, describes a reaction turbine, but it is only a toy in which a ball filled with water is heated over a boiler; when jets of steam issue from the ball, which is mounted on pivots, it spins. But Hero's *Pneumatics* contains a most interesting theoretical introduction on the physical nature of air; Hero does not claim to have invented the theory he describes, but his exposition is excellent. Here he explains the compressibility of air in terms of its structure, adopting a modified atomism. He imagines air to consist of minute particles between which are "disseminated" small amounts of vacuum; an extended vacuum can exist, he believes, only by the exertion of some force, but discontinuous vacua can and do exist continuously between the particles of bodies. These small vacua explain transparency (they allow light to penetrate apparently solid bodies) as well as compressibility; and Hero triumphantly points to the mutual diffusion of two liquids (as wine and water) as confirmation from experience. But though he thoroughly understands the action of syphons, he cannot understand what force causes the water to rise, though he correctly rejects attraction. Further, he knows (as Empedocles had known long before) that when, after filling it, you stop the mouth on top of a spherical vessel whose bottom is a pierced plate, the contained liquid will not flow out. Hero attributes this to the fact that no air can get in "to supply the vacuum," though as soon as air is allowed in, the liquid will flow out. Though the mechanical contrivances described are generally trivial, the explanations of how and why they work are lucid and convincing.

Like pneumatics, optics in the Hellenistic period was both a mathematical and an experimental science. The Pythagoreans were responsible for the notion that vision is caused by something emanating from the eye which falls upon the object—a view long held, and one which readily explains the effect of those possessed of an "evil eye." Not all Greek physicists accepted this: the atomists certainly believed in something corporeal coming from the object to the eye. But the Pythagorean view was accepted by Euclid, whose *Optics* is the earliest extant treatise on mathematical optics. Euclid proceeded in the customary fashion, basing his propositions on certain well-defined assumptions and ordering his proofs geometrically. Most of his propositions are concerned with problems of perspective (the importance of the theater to Greek life in the fifth century B.C. had long made this a subject of interest). Euclid also wrote a treatise on mirrors—that is, on problems of reflection—now lost; in the *Optics* he shows himself familiar with the law of reflection, which had probably been known before his time. There was a long tradition of interest in mirrors (made of polished metal) and of burning-glasses, in the construction of which Archimedes was said to have displayed great skill. Hero wrote on the optics of mirrors, and introduced a mathematical proof of the law of reflection, based on the principle that light always moves in straight lines, the shortest distance between two points. The most important treatise on refraction is that of Ptolemy. He discusses the law of refraction; while he did not discover the sine law, he did have a better approximation than the common assumption that it was identical with that for reflection. Ptolemy performed a number of experiments on the way in which light is refracted in passing from air to water, air to glass, and water to glass, and from these extrapolated to further values (not always with complete success). Ptolemy also discussed atmospheric refraction. Perhaps curiously, the rainbow continued to be interpreted as the result of the pure reflection of light from a cloud, as it had been by Aristotle. But in spite of errors, Greek writers on optics were clear-sighted and able; by the time of Ptolemy both the basic problems and the basic methods of optics had been so clearly stated that optics was to be the best-developed physical science of the Middle Ages.

Chapter 4

THE END OF THE ANCIENT WORLD

Greek science reached its highest level in the centuries following the death of Aristotle; these happen to coincide with the historical period known as the Hellenistic Age. Historians with an eye to literary developments long ago pronounced the age a period of decline; they tended to regard its art—with its insistence on naturalism and emotion—as a decay of the noble principles of classic calm found in the statuary and architecture of fifth-century Athens, and its literary forms as derivative and uninspired. There is no doubt, however, that the third century B.C.—the century of Euclid, Aristarchos, Archimedes, Eratosthenes—was one of the most brilliant periods for Greek science, and that the golden age of science is the Hellenistic period generally. No sign of decline is evident until the beginning of the Christian era: a period coincident with the domination by Rome of the Greek-speaking world.

The Hellenistic Age saw the separation of science and philosophy: as far as we know most of the great scientists displayed little interest in anything but natural philosophy, often confined to narrow limits. Yet at the same time there developed certain personal philosophies—designed to offer a man certainty in the great cosmopolis which lay beyond the tiny Greek cities of the pre-Hellenistic period—which drew heavily on scientific or pseudoscientific doctrine. The two most important of these philosophies were Epicureanism and

Stoicism, both to be enormously popular among the ruling classes of the Roman Empire.

Epicuros was a contemporary of Euclid; about 300 B.C. he settled in Athens to teach his doctrines. He considered that men's primary cause of uneasiness in a troubled world was fear of death, and fear of death in turn stemmed from belief in an afterworld where the gods sat in judgment upon men's souls and condemned them to eternal misery or eternal glory. Epicuros' answer to this was to deny both the existence of life after death and the idea that the gods were interested in men's actions. He was not a true atheist, but he could not imagine a divinity who could be interested in mortal affairs. Epicuros believed that the world was governed by chance, derived from the fortuitous concourse of atoms. He held that the atoms fall normally in straight lines; occasionally and at random they swerve. This swerve accounts both for the meeting of atoms that forms the universe, and for the unpredictability of things that insures freedom of the human will and the occurrence of chance events. Intelligent men, he believed, recognizing all this, would endeavor to live in harmony with the universe by acting with the utmost moral rectitude and adding as little pain as possible to the existing miseries of the world. Epicureanism was turned into a great cosmological system (somewhat reminiscent of the cosmologies of the pre-Socratics) by the Roman poet Lucretius (d. 55 B.C.). His poem *De rerum natura* (*"On the Nature of Things,"* the Latin equivalent of many pre-Socratic Greek treatises) was popular in the early Roman Empire; after a period of neglect in the Christian Middle Ages it was to be popular again in the more pagan fifteenth and sixteenth centuries.

Stoicism, like Epicureanism, was designed to provide a guide to conduct. The Stoics thought of the world as governed, not by chance, but by immutable natural law. Further, they returned to the earlier view of the unity of the cosmos. From these considerations, combined with the belief that a man can better live in harmony with a disastrous world if he knows what to expect, Stoics laid great stress upon astrology (and divination generally). To them it seemed utterly reasonable that what happened in the stars should relate to what happened upon Earth, and what happened in the macrocosmos (the universe) should influence the microcosmos (man). Stoicism is a kind of perversion of science, not unlike in that respect the much later Neoplatonism which was to make number mysticism the key to mystic experience. Stoicism also stressed the concept of the universe as a continuum, in direct contradiction to the doctrine of atomism. Both the idea of a continuum and the idea of natural law persisted; some have seen in Stoicism the vehicle which brought these ideas into modern science.

During the Hellenistic period, the Romans were little interested in science. They were a practical, businesslike, efficient people, magnificent soldiers and administrators, who gradually conquered first the Italian peninsula (including the Greek cities of the south), then Greek Sicily (the death of Archimedes in 212 B.C. was an accidental by-product of the capture of Syracuse), then the western Mediterranean shores. Caesar in the first century B.C. pointed the way to further conquest: north to Gaul and Britain, east to Greece, southeast to Egypt; and in the next century the whole Mediterranean world was brought within the Roman Empire.

Intellectually this changed things little. Greek remained the language of the eastern part of the Empire, and even in the western part Greek was the language of learning. Though old-fashioned Romans at first complained, intellectual Romans soaked up Greek culture, often reproducing it, like Lucretius or Cicero, in Latin garb. Greek slaves were the tutors of rich Roman boys, Athens the city of culture and pleasure, Alexandria the city of advanced learning. But the Romans never lost their belief that much speculation about the world of nature was a waste of time and too apt to distract a man from more serious things. Romans wrote books on agriculture, derived from intimate knowledge; on machinery, like Vitruvius, whose *On Architecture* (late first century B.C.) derives mainly from Greek sources; but above all they liked encyclopedias, handy compendiums which would provide easy digests of Greek knowledge. One of the most influential was that by Varro (c. 50 B.C.), a tremendous editor; his *Nine Books on the Disciplines* contained summaries of the seven liberal arts (grammar, rhetoric, logic, geometry, music, arithmetic, and astronomy) as well as of medicine and architecture; Martianus Capella (c. A.D. 400) drew heavily on Varro. Similarly Celsus (first century A.D.) wrote extensively on a variety of subjects. It is curious that only his treatise on medicine has survived; it is very probably a paraphrase of a Greek work. The Romans did not themselves practice medicine; this explains why the greatest of the late Greek physicians, Galen (A.D. 129–199; cf. Chapter 8), lived for most of his life in Rome, where he had a vast fashionable practice (he was also physician to the gladiators), though he noted wistfully how much better the facilities for research were in the Greek east than they were at Rome. Galen wrote prolifically, always in Greek, since the audience for his scientific treatises was Greek-speaking.

Greatest and deservedly most famous of all the Roman encyclopedists was Pliny the Elder, who died observing the disastrous eruption of Vesuvius in 79 A.D. Pliny's *Natural History* is a vast compilation based upon a lifetime of omnivorous reading. He claimed to have treated twenty thousand topics; not all of the work has survived, but in what remains we have summary accounts of astronomy, geography, mechanical inventions, animals, plants, medical remedies, fishes, metallurgy, and gems, all filled with facts. It is difficult to assess Pliny's mind. Sometimes he appears independent in judgment and rationally critical, as when he derides those who worry about how men stay on the Earth at the antipodes. (He had no doubt whatever of the sphericity of the Earth.) At other times, especially when he describes the mythical behavior of shellfish which wax and wane with the moon, or the sucking fish which is very small but has such an adhesive mouth that it can by mere adhesion prevent a ship from moving, he betrays the crudest form of credulity. He had no ax to grind, but equally he simply could not tell the difference between established fact and fairy tale. And his work is immensely readable—dangerously so to a later age cut off from true Greek science. For the bilingual talent of Roman intellectuals insured that there should be very few translations of Greek scientific texts. This was to have a most unfortunate influence. When, after the collapse of the western Roman Empire, Europe became a purely Latin-speaking region with no access to the Greek language, the intellectual fare offered was poor indeed.

This was, in part, the direct result of a growing anti-intellectualism. The Roman had always looked askance at the Greek intellectual, clever but volatile and unable to govern and administer. There is a famous correspondence between Pliny the Younger—nephew of the encyclopedist, and governor of an eastern province—and the Emperor Trajan, concerned with the building of a water supply for the capital city: to the Roman administrators, the Greek citizens appeared idle children, irresponsibly unable to manage their affairs, though in the end it was a Greek who was to design and build the necessary aqueduct. Indeed, all engineering skills, like all medical skills, were provided by Greeks in the

Roman Empire; the Romans supplied merely the organizing ability. This they valued; they despised the abstract scientific speculation in which Greek natural philosophers delighted; to the Romans, Greek astronomy was only useful in so far as it produced an accurate calendar, and the Julian calendar was only put into effect by the administrative talents of Julius Caesar. (It was to last unchallenged until 1582; it was used in England and America until 1752.) To think rather than to do, to speculate rather than to apply knowledge, all this seemed utterly alien to the Roman spirit.

No wonder, then, that the Latin fathers of the Christian Church were also opposed to scientific speculation. In the first three centuries of the Christian era, when the Roman Empire seemed so strong, there was already abroad a spirit utterly alien to intellectual endeavor—a spirit of mysticism and of despair. As early as the first century A.D., during the reign of the great emperor Augustus, some Romans were predicting that the decay of the old Roman virtues meant death and disaster to all the world. This is a not uncommon view for a displaced aristocracy to hold; the peculiar feature of the Roman case is that the view became so widespread. This was the mood which fostered the growth of numerous mystery religions (where the aim was salvation of the soul for the afterworld through initiation into a secret, closed group). To the Romans, Christianity was but one of these religions, which ranged from Manichaeism, with its vision of the eternal struggle of good and evil, to alchemy, which sought salvation through mystical chemical operations, and which was born in Alexandria about the first century A.D. The way to tranquillity was no longer to live at peace with society; it was to live so that one's soul might find eternal bliss hereafter. Reason could not help; only faith: hence Tertullian's "I believe, because it is impossible"—impossible to the reason. St. Augustine, most intellectual of all the Latin Church fathers, knew the intellect to be a snare, for reason would not let him believe and it was only when he abandoned reason for faith that he found salvation. He permitted only a little science to help in understanding the Bible, with its references to the natural history of the Levant, some simple astronomy to aid in reckoning the Church calendar, and a weakened Platonism to assist in comprehending divine perfection.

Denunciation of reason was more common among the Latin than the Greek fathers of the Church (though it was by no means unknown in the East). Mostly this must be attributed to the difference in intellectual climate between the eastern and western halves of the Empire. For while the West was producing ever more inferior encyclopedias, each more remote from the original sources than the one before, the East produced works of quite high caliber. Not that the Greek world totally escaped the spirit of decline. Both Hero in the first century A.D. and Ptolemy in the second century proclaim this plainly enough, commenting plaintively that of course they cannot be great thinkers, but that it ought to be possible to produce something unknown to "the ancients," the Greeks of the fourth, third, and second centuries B.C., already established as the fountainhead of all wisdom. That so powerful a mind as Ptolemy's could reason thus deplorably demonstrates the temper of the times.

Inability to believe in the possibility of the emergence of new ideas led to useful results in the form of histories and commentaries—invaluable to us now. Pappos (c. A.D. 300) compiled a *Mathematical Collection,* a systematic account of higher mathematics and mechanics, with several original contributions, and Proclos (410–485), who taught at the Academy in Athens, wrote a commentary on Euclid, a mixture of history and philosophy of mathematics with analysis of mathematical problems. Proclos was also the author of an *Outline of Astronomical Hypotheses;* this is an introduction to the work of Hipparchos and Ptolemy, with interesting mathematical details. The most important of these late commentaries—though all were to be of immense value in preserving Greek thought for the benefit of medieval and Renaissance scholars—were those on the works of Aristotle. The treatises of Aristotle as we know them remained in private hands until the first century B.C., when they were edited and published, just at the time when the tradition that wisdom lay in the past first emerged.

Very many commentaries appeared; the most important were those written in the sixth century A.D. by Simplicios and by John Philoponos. The commentary of Simplicios was very widely read in thirteenth- and fourteenth-century Europe, and his views on what Aristotle meant were often accepted as completely authoritative. He also provided a detailed discussion of the astronomical system of concentric spheres, and noted the historical development of various aspects of Aristotelian science. Philoponos was more original and less inclined to agree with Aristotle than Simplicios; his works were less well known, but his ideas on motion did (possibly indirectly) have a marked effect on late-medieval physics.

Philoponos rejected a number of Aristotle's views on motion, notably the explanation of forced motion. He denied that the medium was responsible for the continuation of motion after the initial impulse; on the contrary he believed that as fire imparts heat to a bar of iron, and the heat remains in the iron for some time after it leaves the fire, so a projectile acquires something called *impetus,* which also lasts for some time. And just as the hotter the fire the longer the heat lasts, so the greater the thrust the greater the *impetus,* and the farther the projectile will move. Philoponos also criticized Aristotle's insistence on the necessity of a medium for motion, and argued that motion is possible in a vacuum; his arguments are highly ingenious and intelligent, like all his discussion of moving bodies. But it was the *impetus* theory which was to survive.

That the scientific impulse, though in a somewhat debased form, remained alive among the Greeks of the eastern half of the Roman Empire—the region soon to be known as Byzantium—was fortunate indeed. Though the ancient glory of Greek science lay long in the past, the tradition remained strong. The very respect for "the ancients" which prevented men from making new contributions led them to try to preserve the knowledge of the past. It was to be many centuries before scientists anywhere in the world could make any great contribution to science, or reestablish the tradition of scientific progress. But the Greek scientists of the Christian era had ensured that knowledge was preserved, locked away in the Greek tongue to be sure, but available to any who might want it enough to make the effort of translation.

II

Philosophy and Physics in Medieval Europe

Chapter 5

EUROPE REDISCOVERS ITS PAST

Romulus Augustus [Augustulus], deposed by a barbarian in 476, was the last of the authentic Roman Emperors in the West. More than three centuries elapsed before the Imperial crown was placed upon the Frank, Charlemagne, by the hands of Pope Leo III on Christmas Day, 800. It was easier to revive the name than the reality. Even in its fifth-century decadence Rome had not renounced its civilization; its cities still stood in Italy, France, Spain, and Britain; men like Macrobius and Martianus Capella, who loved books and learning, strove pathetically to prevent the total loss of their own already degenerate knowledge. But the next three centuries had seen the sweep over all Europe of the still barbarous Germanic peoples, the destruction of towns and commerce, and the resignation of the last vestiges of ancient thought to a handful of semiliterate priests and monks. Only one point of cultural stability endured; upon the papacy in Rome rested virtually the whole responsibility for restoring civilization to Europe. Greek philosophy and science were totally cut off; of the wreck only one major work, Plato's *Timaeus,* was salvaged in Latin translation. As Christianity pushed north, east, and west, however, bearing with it the Bible and the tomes of the Fathers of the Church, medieval learning began to assume its characteristic form: truncated, feeding on the remnants of a more glorious (yet distrusted) past, clerical, and always subject to Christian dogma.

Charlemagne, Alfred of England, and their learned bishops encouraged a safe and pious exercise, not an intellectual adventure.

Yet this is what, despite authority, the recovery of a lost inheritance became. In a world where literacy was rare and superstition universal, a few men had begun to put together new books. In the sixth century Boethius compiled from Aristotle, Euclid, and other classical sources long-used manuals of logic, astronomy, and mathematics, and Alexander of Tralles (a Greek) wrote on medicine at Rome. In the seventh century Isidore of Seville created in his *Etymologies* the model of many medieval compendia of knowledge. In the eighth century Bede proved that Pliny and arithmetic were not unknown in the far north, while his countryman, Alcuin, was the chief architect of the schools in the "Carolingian revival." Yet it was a thin thread of learning that persisted, when geometry consisted of the mere enunciation of a few of Euclid's theorems without the demonstrations, and scholars could argue about the meaning of the word *parallelogram!* Naivety, too often, had taken the place of the inquiring mind and mythology the place of science. To break out of this prison of ignorance, to trace the backwards path and, from the few broken pieces, to restore the mirror reflecting the brilliance of antiquity, was indeed an adventure.

At this point, unquestionably, the intellectual life of China was far advanced beyond that of western Europe. Nearer home, too, in other societies learning was deeper and criticism sharper. Constantinople, impregnable until it fell before the western Christians in 1204, remained the capital of a sizable territory that had been the nucleus of the eastern Roman Empire. Though strained by incessant attacks on its borders, Byzantium had never been swept by barbarians nor parceled into chieftainries; in unbroken tradition it had preserved the Greek language and Greek science. At a small distance, yet virtually inaccessible, was almost all that the Latin West was striving to recapture. If Byzantium had not improved much on its legacy, at least it had maintained it. Aristotle, Galen, Euclid, and Ptolemy were not forgotten; their works still existed, if the West could only learn to read them. Byzantium was not merely a museum, either; it showed its practical ingenuity in the earliest form of chemical warfare (the pyrotechnic weapon known as "Greek fire") and its intellectual ingenuity in such philosophers as John Philoponos, author of the first shadow of one of the central concepts of modern science (pp. 56–57).

Distance and the theological gulf between Catholic West and Orthodox East always impeded a free cultural exchange between the two halves of the Christian world; linguistic ignorance and poverty of intellectual development rendered Europe in Charlemagne's time, and long after, quite unable to grasp what it was missing and longed for. The first great influence, the first resource for stocking the barren shelves of monastic libraries, was found in another quarter, nearer yet stranger. Already in the eighth century the followers of Mohammed, having poured westwards across North Africa and crossed the straits into Spain, had been repulsed on the very margin of France. At the other extreme of its empire Islam had thrust northwards and eastwards, through Syria, Mesopotamia, and Persia into northern India, occupying all the eastern half of the civilized world except the remnant that was Byzantium. There was inevitably devastation (the Arab, it is said, makes the desert) as there was intolerance; but many rich communities and even Christian monasteries survived under the new regime.* The rapid spread of Islam did not effect such a total collapse of a thin veneer of civilization as occurred in Europe with the barbarian invasions—partly because civilized life, crafts, and learning had evolved in the east gradually over thousands of years. Moreover, although Islam was founded in Arabia (its calendar begins in A.D. 622, the year of Muhammad's withdrawal from Mecca to Medina with his first disciples), the conquering rulers were by no means all wild Bedouin of the desert. Its language was the most clearly Arabic feature of the new empire, which was spread by proselytism rather than by violence. Syrians, Persians, Egyptians, Jews, Berbers, "Spaniards," Sicilians, and even Italians became Muslims and speakers of Arabic—the only language of their new faith, since the Koran is profaned by translation. With astonishing rapidity Arabic became the greatest vehicle of science, learning, and literature in the world. Far from shunning the delights of civilization the religion of the desert gave them passionate encouragement; it sponsored some of the most beautiful buildings the world has ever seen, the exquisite miniature art of Persia, the brilliant faience, fine metalwork and other crafts, which the West too was to borrow in the end. Within the wide scope of Islam life recovered, for the

* The greatest loss to science attributed to the Islamic conquests was the burning of the Library of the Museum at Alexandria, in 640.

few, an elegance and leisure that Greece had never known and Rome at its height had scarcely equaled.

Intellectually it was to Greece and Rome that Islam—like Europe—looked for its inheritance. But what a difference! Islam, from its first conquests, was rich in learned men; the treasures of antiquity remained in Greek-speaking lands that Rome, in her time, had governed. Moreover, much of the ancient philosophy and science had already been translated into Syriac (Aramaic) or Persian, both languages accessible to Muslims. Such work had been going on strenuously at Nisibis in Syria and Jundishapur in Persia long before the Arabic conquest, the translators being polyglot Nestorian Christians. Within a century of the Prophet's death the Muslims were seizing eagerly upon this literature, the two great Abbasid Caliphs al-Mansur (the founder of Bagdad, d. 775) and Harun al-Rashid (763–809, hero of the *Arabian Nights*) being munificent patrons of translation into Arabic from Syriac, Persian, Greek, and even Hindi. Under their stimulus Bagdad became the most mentally vigorous city in the world, and it was not long before other lesser centers of learning graced the whole Islamic empire.

That empire remained, as it was created, an international enterprise; although held together by some common cultural traditions, by a common official language (counteracting the divisiveness of its many national tongues) and by a common religion (which yet tolerated Jews and Christians), it was divided into numerous political and religious factions. Arabic science too was an international enterprise, for by no means all who contributed to its rise were Arabs or even Muslims. Avicenna (980–1037), the greatest Arabic scientist, was a Persian and sometimes wrote in his native tongue. Averroës (1126–1198), the greatest of Islamic philosophers, was born at Córdova in Spain and spent most of his life there. Mashallah (d. c. 820), one of the earliest Arabic astronomers and author of a treatise on the astrolabe that was translated into English by Chaucer centuries later, was a Jew like many others who wrote in Arabic after him.* Hunain ibn Ishaq (810–877), foremost translator of Greek medical works into Arabic, who

* The astrolabe, one of the favorite medieval astronomical instruments, consists of flat, engraved plates representing a projection of the celestial sphere. It demonstrates for any given day, without calculation, the position of the sun among the stars, the hours of rising and setting of the sun and stars, the time, the meridian altitude of a sun or star, and so on. Its construction requires accurate observation and fairly complex geometry.

came from Jundishapur to Bagdad, was a Christian and so presumably was his son, Yaqub Ishaq, who translated a number of mathematical and astronomical books. Every region and every nation of Islam contributed its quota to the development of science, and ultimately to the intellectual life of medieval Europe. Islam also drew upon the mathematical and astronomical knowledge of India, an obvious example being the borrowing of Hindu numerals, which became the "Arabic" numerals of Europe.

This catholicity was certainly one factor in the sudden flowering of Arabic science, which it is not easy to account for. Certainly Islam occupied a rich region of the globe (agriculturally far richer then than it is now), and controlled a flourishing commerce; its religion, too, favored literacy and learning. But perhaps the most important reason why science particularly flourished was the support given to it by the caliphs and other rulers. The caliph al-Mamun (786–833), for example, followed the earlier Abbasid caliphs in encouraging science at Bagdad, sending a mission to Byzantium to obtain fresh manuscripts and founding a House of Wisdom in his capital, a scientific institution that was equipped with an astronomical observatory. In Cairo, too, was a similar academy, founded in 966, supported by the astronomer-caliph al-Hakim. Here there was an observatory where worked some of the most famous astronomers of Islam and a man who was outstandingly its greatest physicist, Alhazen (c. 965–1039; p. 81). The contemporary caliph of Córdova, al-Hakam, exercised an equally enlightened patronage and was himself one of the greatest medieval scholars; his library, it is said, numbered nearly half a million volumes. Other Spanish cities—Toledo, Seville, Granada—were only less important centers for philosophy and science while they remained under Muslim rule. More than vestiges of this intellectual eminence survived after the Christian reconquest, as was equally true in Palermo, capital first of Muslim and then of Norman Sicily.

Thus Christian Europe, whose southern and western fringes were actually part of the Islamic empire, was necessarily thrust against a rival society towards which it felt the strongest hostility, yet whose superiority in philosophy, science, and technology it was gradually forced to recognize. Not until the tenth and eleventh centuries did Latin Christendom begin to be dimly aware that it shared, or might seek to share, a common intellectual heritage with Islam; not until the twelfth century were the first efforts made to render this heritage accessible in Latin. The occasion was provided by the slow retreat of Islam; Christians became willing to learn from Muslims and Jews only when they were indisputably their masters. Hence, although Christendom and Islam shared the intimacy of war during the several centuries of the Crusades, and although Latin kingdoms were established in Syria and Palestine, learning gained little from this quarter because the Christians were not the masters.

After the final check to the Muslim advance in the eighth century the little Christian principalities in the north of Spain expanded to the south and north, until at last Toledo was captured in 1085. Necessarily they tolerated Muslims and Jews among their new subjects; the schools, libraries, and scholars of the former power were now in Christian hands. The ideal setting for cultural borrowing was created, as it was also in Sicily by similar violent means. Early in the eleventh century adventuring Norman barons drove the last relics of Byzantium from the south of Italy; this done, they looked across the sea and by 1091 completed the conquest of Muslim Sicily also. There they maintained for a time a considerable kingdom, half Norman, half Muslim, with a court that regarded translators, scholars, and physicians benevolently. In medicine especially, through the "School" (later University) of Salerno, this region of Europe where Latin, Greek, and Arabic were brought into such close proximity was to exercise a formative influence.

Embassies had long been exchanged between Christian and Muslim rulers, and works written long before the twelfth century reveal Islamic influence. But there is no bulk of known translation from Arabic prior to that of an Englishman named Adelard of Bath, who during the early part of the twelfth century traveled in Italy, Sicily, and the Near East; he translated Euclid's *Elements* from an Arabic version, and several other works by Arabic mathematicians. He wrote books of his own utilizing his wide reading among new manuscripts, the most influential being *Natural Questions*. The next and more voluminous group of translations are all associated with Spain; they were made by one John, at Seville, by Herman of Carinthia (who was in Spain 1138–1142), by Robert of Chester, and by Plato of Tivoli, who lived in Barcelona 1134–1145. The most successful of all the translators from Arabic, Gerard of Cremona, worked at Toledo for many years and died there in 1187. Although about seventy translations are attributed to him it is clear that these cannot all have been the unaided work of one man. Gerard, and probably most of the other early translators, were dependent upon collaborators, in some instances a Jewish scholar who knew both Arabic and Latin. Not surprisingly, therefore, their texts were often imperfect because of their failure to understand the true meaning of the originals, and because of the absence from Latin of the requisite technical terms. Perforce many Arabic words entered that language (still, and long to remain, the universal language of scientists and scholars) and through Latin the European tongues.

Just as Arabic literature had been vastly enriched by translation a few centuries before, so was Latin literature in the twelfth century—and indeed later, though with a less copious flow. Books relating to every branch of knowledge were translated—mathematics, philosophy, astronomy, geography, alchemy, physics, and every aspect of medicine. No one has ever computed how many learned treatises were thus turned into Latin during the Middle Ages—perhaps between two and five thousand. The effect was as though the whole of Europe had been put to school in Islam. For not merely were the master works of Greece translated; the books of the Muslims written since the birth of their learning in the eighth century were translated in even larger numbers.* The science which Europe adopted avidly in the twelfth and thirteenth centuries was not the science of antiquity; it was Arabic science, with the weaknesses and strengths that four centuries of thought and observation in Islam had grafted on the Greek foundation.

The extent of this Arabicization was variable, of course. Alchemy, as its name implies, was almost totally Arabic until some Byzantine chemical skill came in later. Aristotle, on the other hand, was studied by medieval Europeans pretty much in the Greek tradition, although the texts used were sometimes derived from Islam and although Muslim philosophers had a profound influence upon the growth of the medieval tradition of philosophy. In astronomy and in medicine the authority of the Arabs was very great and the original Greek founders, like Ptolemy and Galen, were relatively little read. In mathematics again, while geometry was studied in Euclid, Archimedes and the later Greek mathematicians were practically unknown and European mathematics was much guided by Islamic studies in arithmetic and algebra (another Arabic word). In short, Europe tried to

* The Arabs, for obvious reasons, had little access to the Roman writers, like Lucretius, who were also virtually unknown in the western tradition.

assimilate *all* that had gone before, often without due critical assessment of its worth. Some parts of the Arabic literature, infected by the false religion of Islam (as Christians saw it) were rejected outright. But for the most part Latin scholars tended to treat all their predecessors with excessive respect, as "authorities." They scarcely observed the fact that more than a thousand years separated Aristotle and Avicenna, nor did they all reckon that as the Arabs might have known the Greeks imperfectly, or misinterpreted them, so they themselves lay open to errors in their own borrowings from the Arabs.

To some extent the likelihood of distortion in the transmission of ideas to Europe was lessened by direct translation from Greek into Latin, a conversion increasingly significant towards the end of the Middle Ages. There had, perhaps, never been a time without some knowledge of Greek in Europe nor a complete break in the flow of translations. But it was very thin until the twelfth century, and almost without cumulative effect. For example, the more advanced logical works of Aristotle had been translated long before by Boethius, but their explosive effect followed from a new translation out of the Greek by James of Venice, about 1130. Many works, like these (which were also rendered from Arabic by Gerard of Cremona), came into Europe twice, once direct from Greek, once through Arabic. Either might have the greater currency: Ptolemy's *Almagest* was first translated into Latin from Greek in Sicily, about 1160, but it was the version from Arabic by Gerard (about 1175) that was used in medieval Europe. In the next century, when the available texts were being scrutinized rather more narrowly, some retranslations were made from Greek specifically in the hope of better recovering the meaning of the original; most active in this respect was the Flemish translator William of Moerbeke, who was guided by his friend Thomas Aquinas in choosing his subjects.

It is possible that but for the translators Europe might never have become aware of its intellectual heritage, although this was rendered unlikely by the Christianization of the barbarians. More plausibly, Europe might have rediscovered its past solely through a gradual awakening of interest in neglected Greek and Latin manuscripts. In the event, however, Europe was heavily indebted to Islam for its intellectual endowment, to such an extent that the Middle Ages never quite formed their own independent picture of ancient thought. The mirror held by Avicenna, Rhazes, al-Biruni, al-

Kwarizmi, and so many more, by which the brilliance of Greek science was reflected to the Middle Ages, was broadly true, but there were some things it suppressed, some things whose emphasis it shifted. Europe found its past in that mirror, and set to work to understand it, to assimilate it, and to systematize it. Scholasticism was brought into being by men who used logic as an intellectual weapon with a confidence that has never been surpassed; others turned to optics, mathematics, astronomy, alchemy. Rising from semibarbarism, western Christendom was able to present, after three centuries (1050–1350), the three centers in the world most stimulating to the intellect—Bologna, Oxford, and Paris. It was a fantastic cultural achievement, greater even than that which Islam had effected. Yet, despite all this, medieval Europe was still remote from Greece.

Medieval science had just reached its peak when, at the end of the fourteenth century, this fact became apparent. There were already by this time men (writers, not scholars) who, although by no means critical of contemporary learning, were pursuing a path independent of it. Echoes of medieval philosophy and science are assuredly to be found in the poetry of Chaucer and Dante, for these defined the cosmos in which the poets moved upon their own purposes. But their purposes were not the scholars', and in Chaucer at least there is a strong hint of mockery at pedantic pretensions. So of the alchemists:

Their scientific jargon is so woolly
No one can hope to understand it fully,
Not as intelligence goes nowadays.
And they may go on chattering like jays
And take delight and trouble in their chatter,
But for all that they'll never solve the matter.
If you are rich it's easy to be taught
How to transmute and bring your wealth to naught.[1]

Petrarch, Dante, Chaucer, Boccaccio, Botticelli—theirs was another view of man, of the creative intellect, than that of the scholar. Nobility, misery, sin, beauty, mystery, and simplicity—everything was part of the tragicomedy of human experience which was after all the sum of human knowledge. Nor was man only a creature of emotion; he was the political animal, a decisive figure on the world's stage. He had created society and its history. All this the late fourteenth century began to perceive, reacting against logic, reacting against the unhappiness of a tragic time (for the Black Death was a recent memory, and the Hundred Years' War a present misery), reacting against the papal predominance in all things. And these uneasy rather than dissident spirits found—as the Middle Ages always found—a fresh model for their own aspirations in antiquity. Some of the imaginative literature of antiquity, the poetry of Horace and Vergil, for example, was well known; much that had been long forgotten was now rediscovered, along with drama, history, and political writings.

The Latin literature was readily accessible, once it was brought to light, as (to mention only two scientific classics) Lucretius' didactic poem *On the Nature of Things* was brought to light in 1417, and Paracelsus' *On Medicine* in 1426. Such new discoveries rapidly assumed an importance even greater than their true intellectual merits deserved. But the real key was Greek, and now was born the conviction that dominated education for centuries: Latin commanded the essentials of knowledge, Greek its refinement and delight. So it happened that when a certain Byzantine, Manuel Chrysoloras, began to teach his tongue in Florence in 1397, his lectures found a rapturous audience. The moment was opportune. What was left of Byzantium was no longer remote; Europe traded there regularly. Its defeats rendered it an object of fear no longer, and it was too dependent on the West to be proud. As from Spain centuries before there was a steady flow of manuscripts, borne to Europe in triumph, that ceased only when Constantinople fell to the Turks in 1453. At last European learning was directly in touch with the finest minds of the ancient world, and the printing press was ready to disseminate this renaissance of learning more swiftly than ever thoughts had traveled before.

So far as science was concerned, for about a century the main task was to be the making of editions, translations, commentaries, and extensions of what the Greeks had achieved. In a sense it was the twelfth century all over again, but at a far higher level. For if the scientist and scholar of the Renaissance looked back to a Golden Age in Greece, medieval thought had prepared him to appreciate, and even emulate, the Greek achievement. To the men of the Renaissance the Middle Ages seemed blind and narrow, excessively preoccupied with the minutiae of theology and skilled only in verbal quibbles. Indeed, these neoclassicists, like Erasmus (1465–1536), sometimes dismissed medieval discussions that now seem of the highest importance for the development of scientific thought as meaningless jargon. They condemned the barbarity of scholastic Latin, and above all the scholastic dependence on Arabic writers in place of the Greek originals. As Michael Servetus (p. 125) put it, speaking of Galen:

> In our happy age, he who was once shamefully misunderstood is reborn and re-established to shine in his former lustre; so that like one returning home he has delivered the citadel held by the forces of the Arabs, and has cleansed those things which had been befouled by the sordid corruptions of the barbarians.[2]

On this view it was essential to knowledge that scholars should begin all over again with the purest Greek and Latin texts, ignoring all that Islam and Europe had accomplished. In practice this destructive effort of the Renaissance was only partially successful. For although in medicine, at one extreme, the Arabic authors were almost entirely displaced by the mid-sixteenth century, the discussion of certain physical problems continued in the same unbroken pattern from the fourteenth century to Galileo.

The Renaissance revulsion from the Middle Ages was, of course, exaggerated. The neoclassicist position was at times absurd—when Machiavelli argued that the weapons of the Roman legion were superior to contemporary firearms, for example—and always open to the criticism that it led merely to imitation and a narrowness of outlook no less marked than that of scholasticism. It would have been a tragedy if men had really acted upon the belief that the ancients had reached the possible limits of knowledge; fortunately, they did not. But, though the work of discovering and editing texts detracted from original inquiry, and though dogmatism might result (as in the belief that Galen was never in error), the neoclassicism of the Renaissance did prove fruitful. In the medical sciences and in mathematics the Greeks had done work that far excelled that of the Middle Ages. In astronomy, too, it was profitable to learn thoroughly what Ptolemy's geometry of the heavens really had been. And so on; but above all else by reverting to the Greeks the Renaissance opened up new perspectives for thought: atomistic speculation by the pre-Socratic philosophers and Lucretius; the mathematical method in science used by Archimedes; the biological theories of Aristotle. The rigidity of the Arabic-medieval picture of ancient thought was broken; the Renaissance learned that many *different* kinds of scientific procedure and idea are possible. When Copernicus became thoroughly dissatisfied with the Ptolemaic account of astronomy (he tells us) he looked through the ancient authors to see if any had ever proceeded in an alternative fashion; when he found that they had, he began to investigate their ideas. Thus even the neoclassicists learned that science is not dogma, that there are always alternative hypotheses which may be tested by their results.

By the mid-sixteenth century Europe was fully in possession of its intellectual history. In future it was to gain little except wisdom by looking backwards. Already physicians, anatomists, mathematicians, astronomers, physicists, were attacking unsolved problems, finding their bases in Greek science and some of their strong resources in the Middle Ages. They were already beginning to think that the Golden Age for science lay not anywhere in the past but in the future, created by men's continued efforts.

Chapter 6

THE RISE AND FALL OF ARISTOTLE

There were three learned professions in the Middle Ages: the Church, the law, and medicine. In the later Middle Ages candidates for each of them were commonly (though not invariably) educated at the universities, which first acquired their peculiar corporate character in the twelfth century.* Their origin was in cathedral or monastic schools, whose elevation to an advanced level of instruction in universal knowledge *(studium generale)* was a direct result of the translating fervor and of the cultural renaissance that followed from it. Hence the university was a clerical institution, even when its studies included civil law and medicine; the greater number of the teachers were clergy, and most of the students were destined to take orders.

Upon this body of men and boys—upon the teachers and the students who would go out into the world—the responsibility for maintaining the intellectual life of Europe rested. For outside these three professions the ranks of the literate were thin indeed. Of the sciences, only medicine and surgery were cultivated by laymen, because the medical profession was at least nominally barred to priests; all else fell to the

* In England civil lawyers trained in separate corporations, the Inns of Court. There were, inevitably, many ill-educated priests and physicians for want of better, while surgeons, barber-surgeons, and apothecaries had rarely received any liberal education. These were semiliterate craftsmen.

Church and was perforce accommodated to an educational system whose highest aim was to teach true knowledge of religion. Long before, the Fathers of the Church had expressed fearful reservations about pagan skill in astronomy and philosophy, and many another good Christian had thought that it was a sufficient task to learn to read the Bible and to know how to save one's soul. Yet always—as since then—there were others who opposed obscurantism, and used the argument which has brought credit to science down the centuries: God created the universe, and it is proper to know his works in order to admire his majesty. It is hard to suppress curiosity utterly, and always possible to justify it. Though fears tormented the superstitious, in the minds of learned men the end of the world no longer seemed imminent; they, and their church, gradually became adapted to living the good life in this world. Ignorance in a priest or monk became a shame, if not a sin, and more stress was laid on the importance of astronomy for keeping the Christian calendar correct, on medicine as a work of charity, and on philosophy as a guide to the true understanding of Scripture. By the fourteenth century, discussion of very abstruse questions in mathematical physics had become, by this devious route, a perfectly proper element in the education of a priest.

The groundwork of medieval education was furnished by the seven liberal arts, of which the more elementary *trivium* (grammar, rhetoric, logic) was concerned with expression, and the more advanced *quadrivium* (arithmetic, geometry, music, astronomy) wholly with mathematics. Upon this the universities also superposed the professional training already mentioned, which for the priest comprised philosophy and theology. In practice, however, teachers were to be found in the universities who were deeply learned in one or more of the various branches of science, or in philosophy, or in medicine, and they were free to lecture as profoundly as they pleased upon their various specialties. For all its clerical context, the medieval university was a place of great (though never complete) freedom and, under the guise of propounding hypotheses or plausible objections, the medieval scholar was free to discuss—though not to affirm as true—almost any imaginable notion about the physical universe.

The mathematical sciences, because of their traditional place in the educational scheme of things, and medicine because of its utility, had a comparatively easy passage when translated knowledge entered Europe during the twelfth century. There was inevitably some feeling that Christians should not sit at the feet of infidels, but this did not last long. In philosophy the resistance against the assimilation of Greek and Islamic thought was more prolonged. Moral philosophy was an especially tender point, because of its close link with religion, but even in the natural philosophy (physical science) that came from Sicily and Spain the theologian could find much that was highly objectionable. Natural philosophy touched on such questions as the formation, age, and duration of the universe; its arrangement (within which room must be found for Heaven and Hell); the physical aspects of the divine plan; the relation of body and soul; or the relationship of the stars to the freedom of the human will. Undirected, the natural philosopher might innocently pursue lines of argument whose ultimate consequences would be damaging to Christian belief; certainly he could not be permitted to hold (with Aristotle) that the universe is uncreated, or (with the Greek atomists) that it is the product of chance coalitions of atoms. Reason was but a fallible guide on such delicate matters; sometimes (the medieval scientist always recognized) faith must be followed even when it proceeded directly contrary to reason.

In modern times, until science escaped from religious circumscription altogether, it was commonly argued that there must be an ultimate reconciliation of the two truths, that of science and that of Christianity, which would be attained whenever scientists knew enough. The medieval scholar was less complacent. He admitted that some religious truths were and must remain unreasonable, like miracles. In a sense, then, he adopted a double standard: in matters to which Christian authority was indifferent, among which most scientific questions were numbered, he followed the dictates of rational inquiry; in matters upon which the Christian could only hold one view, the medieval philosopher taught that view with the strength of argument added to faith. To give one instance: in the late fourteenth century the great French philosopher and scientist Nicole Oresme (1323–1382) discussed the motion of the Earth, treating it as a hypothetical question, as he was free to do. He pointed out that if the Earth were supposed not to be fixed, but to rotate on its axis once every twenty-four hours, all the phenomena of astronomy could be explained as well as they conventionally were by supposing the Earth to be fixed at the very center of the Universe.* He even developed arguments indicating that the

* Oresme did not go so far as to imagine that the Earth has a progressive revolution about the Sun.

Earth's rotation might more plausibly explain the phenomena, and others confuting the obvious objections against such a rotation. In short, Oresme clearly understood that, on balance, to believe in the rotation of the Earth is more reasonable than to believe in its fixity; yet he rejected the former opinion for the decisive reason that the Bible speaks clearly of the Earth's standing still. To Oresme, accordingly, his own arguments for the Earth's rotation were merely specious. Only the direct evidence of revelation saved man from this scientific error into which his unaided reason might have trapped him.

Given the medieval view, that reason (employing the available evidence) *could* yield error, what was the best defense against falling into such error concerning those questions on which God had remained silent? This was the central problem of medieval philosophy and hence the central problem of medieval science, which was always, intellectually speaking, a direct offshoot of philosophy. A familiar modern answer is "make experiments." However, while making experiments is a splendid way of acquiring factual information about definite problems, and in certain cases of testing the validity of hypothetical ideas, it does not by itself provide a method for solving *all* problems at *all* stages of science. No experiments in the Middle Ages could have decided whether the Earth rotates or the stars revolve, nor whether a complete vacuity in Nature is possible, nor whether all animals derive from eggs. Yet there were men at this time, as there had always been, who were passionately interested in questions like these which could not be answered by making experiments (though it is also true that they were interested in others which could, without excessive ingenuity, have been subjected to experiment). Scientific thought would have developed very slowly, and perhaps not at all, if it had grown only with the ability to discover how to make experiments fruitfully.

Not that the Middle Ages wholly neglected experimentation, as we shall see later. But in fact the medieval natural philosopher was for the most part far less concerned with the details of the universe than with general ideas of the universe. This attitude is highly repugnant to some modern scientists who regard accurate knowledge of details as the very essence of science, forgetting perhaps that such knowledge is no substitute for general ideas, about which men have a perennial and highly creative curiosity. A medieval philosopher did not regard himself as making early contributions to the modern sciences of astronomy or geology, of which he could have no conception and whose current research problems are significant only to experts; rather he found himself asking such direct, universally interesting questions as "Why does the Sun move?" or "Where did the water in the oceans come from?" He found answers to such questions sometimes in the Bible (which was infallible), more often in the writings of his Greek and Islamic predecessors. These latter answers seemed to him for the most part plausible, but not unquestionably correct. He found that they involved certain concepts, certain logical maneuvers used in conjunction with these concepts, and ultimately certain theories formulated as a result. Accepting this scientific inheritance —and there was assuredly no better starting point available to him—the medieval philosopher began to examine the concepts, the logic, and the theories. He did so by investigating the nature, structure, and implications of each, seeking to discover whether concepts, logic, and structure were adequate to fulfill the function imposed on them, whether they were consistent, and whether they conformed to reason as he understood it and to the world as he knew it.

To some extent the medieval natural philosopher was merely giving a new emphasis to the conceptual character of Greek science, which the Arabs had not modified. Science has two chief functions: firstly to describe the world in which men live, and secondly to explain it, rendering it intelligible by disclosing the manner in which causes operate in nature and in which phenomena are subject to universal laws. The Greeks, too, had been chiefly preoccupied with the discovery of such general concepts as would make nature intelligible, especially those Greeks like Aristotle and Galen whom both Islam and Christendom admired most. The task that dominated medieval scientific thought was the exploration and criticism of these Greek concepts—the idea of circular motion in the heavens, the physiological concept of humors, the Aristotelian concepts of matter. Moreover, the manner in which the Middle Ages adopted Greek science as the legacy of a superior order of men militated against doubt of the facts reported, unless they were obviously mistaken. It was worthwhile to examine Aristotle's or Galen's reasoning, if only to be able to explain it properly and overwhelm objectors, but it was hardly conceivable that their facts could be wrong—especially as, in the premature evolution of general ideas, the facts brought in evidence are usually quite commonplace and indeed correct. As for making a wholly fresh beginning, rejecting such giants as Aristotle and Galen altogether, that (given the historical situation) would have seemed and would indeed have been absurd folly. Finally, the natural philosopher was himself heir to and involved by circumstances in a tradition which dealt with the analysis of concepts and their implications, not with the verification of facts: the theological tradition. In theology the facts were certain, revealed. The expositor of theological doctrine could only explain and comment, comparing one passage with another, removing apparent contradictions, bringing out the force of one doctrine for everyday conduct, or strengthening another by new quotations. The theologian is confined by the tenets of his religion, unless he loses his faith. It is hardly too much to say that the Middle Ages studied science as though it were theology and Aristotle's *Physics* as though it were the Bible. The philosopher's best contributions, like the theologian's, took the form of long footnotes to the sacred text. Until he lost his faith.

The plain fact is that twelfth-century Europe absolutely lacked men equipped to criticize Aristotle's science as a modern scientist would; it did not wholly lack able philosophers, who asked of Aristotle's science not "Is it true?" but rather (as a modern philosopher might) "Are these theories consistent?" and "What do these concepts mean?" The attempt to answer these questions had, by the end of the Middle Ages, produced a vast volume of commentary, and some science; by contrast, rather little had resulted from medieval attempts at direct observation and experiment. What happened with respect to scientific questions was strongly influenced by the effect of the new learning on the premier field of knowledge, theology. For what first captured the enthusiasm of twelfth-century thinkers was the power of the "new logic" (Aristotle's more advanced writings on logic) as an instrument of thought; and this instrument was, naturally, turned first upon theology. The great logician of that age was Peter Abelard (1079–1142), who wrote:

> My rivals have recently devised a new calumny against me because I write upon the dialectic art, affirming that it is not lawful for a Christian to treat of things that do not pertain to the faith. . . . (But) all knowledge is good, even that which relates to evil, because a righteous man must have it.[1]

Abelard scandalized the Church, not by offering arguments against its teachings, but by showing that they were inconsistent, "tangled in difficulties." Though utterly uninterested in science himself, he in fact first used the logical methods of attack employed by later natural philosophers, who both dissected the "tangled difficulties" of scientific theory in his way and sought (like the theologians) to remedy by new syntheses the logical imperfections thus discovered. In brief, Abelard was the first great scholastic, since this procedure is characteristic of scholasticism in all its manifestations.

For every aspect of medieval thought the great synthesis was that of Thomas Aquinas (1225–1274), the outstanding scholastic, who laid the bases of modern Aristotelianism in Europe. He was not the first since the translators to shape the course of scientific speculation (some of his predecessors will be mentioned later), nor was his mind strongly drawn to natural philosophy, except as it was a branch of universal knowledge. He contributed nothing new to physical science but strove to make it a system coherent with philosophy, to which end he prepared commentaries upon four of Aristotle's scientific treatises (including the *Physics*) and himself wrote works with such titles as "On the Principles of Nature" and "The Hidden Operations of Nature." So far as science was concerned, the Thomist synthesis was, naturally, a purely logical synthesis; it did not in any way improve the description of the universe. Thus the great authority of Aquinas as a theologian and philosopher only confirmed the medieval tendency to assimilate natural philosophy to metaphysics and theology, and to render the study of nature a purely intellectual exercise conducted with the appropriate logical tools. Indeed, his singular achievement was to remove the last discrepancies between the Christian world-view and that of Aristotelian science (partly by suppressing or repudiating some of Aristotle's own concepts) so that the development of both could, he thought, proceed in harmonious freedom from the suspicion that the translations had provoked. In this aim Aquinas was not, at first, completely successful: certain of his Aristotelian theses were condemned in 1277 by the Universities of Paris and Oxford, as such theses had been condemned earlier in the century. These condemnations were of no ultimate effect—at least they did not obstruct the ascendancy of Aristotle over European thought—but they may perhaps have reinforced the distinction between the truth of faith and the truth of reason, and so to some extent increased the natural philosophers' freedom to speculate.

Aquinas was the pupil of Albert the Great (1193–1280) who taught for three years at Paris but spent the last years of his life at Cologne. Albert is remarkable not so much for his original contributions to knowledge as for the wide-ranging writings by which he established the new learning securely in Europe. A fervent Aristotelian, he rejected (like all his more orthodox successors) the freer forms of Aristotelian speculation linked with the name of Averroës. Besides Aristotle and many Arabic authors, Albert drew largely from other ancient authors, especially Theophrastos and Pliny, for he was much interested in natural history, being one of the few medieval writers to record original observations upon plants and animals. His works were freely quarried in later times, and indeed he had collected most of the material that was to be found in other medieval encyclopedias, such as the *Speculum* of Vincent of Beauvais (d. 1264), which was much larger and more modern than the long-popular compilation of Bartholomew the Englishman (c. 1235), *On the Properties of Things.*

However far separated in intellectual level, there is a certain affinity between such compilations as these (which remained in current use for the next two centuries and more) and the synthetic works of Aquinas; both are books made out of books, whose authors show a consistent lack of interest in the agreement between books (or ideas) and nature. Aquinas' other great predecessor, the Englishman Robert Grosseteste (c. 1168–1253), added a more developed empiricism to his logical interests. He was concerned not only for the consistency of theories, but for their applicability; moreover, he was himself a translator from Greek, and the refounder of philosophical studies at Oxford. Like all other medieval philosophers he was deeply read in Aristotle and an expert logician; Grosseteste's particular eminence lies in his deep interest in the problem of testing scientific propositions. Only when such propositions were demonstrated as true could the genuine causes of things be known.* Thus, he recognized the formal character of demonstrations in mathematics or mathematical physics, which are derived from appropriate definitions and axioms. However, all propositions

* Grosseteste, being a good Aristotelian, could be satisfied with teleological causes, as when he wrote: "The cause of having horns is not having teeth in both jaws, and not having teeth in both jaws is the cause of having several stomachs."[2]

in science do not have this formal truth, nor do the axioms and definitions of a mathematical science (such as "light is always propagated in straight lines," the basic axiom of geometrical optics). Accordingly, Grosseteste argued that such propositions should be verified by reference to experience (not quite yet, in the modern connotation, experiment). This had been urged by Aristotle and Galen long before, though often neglected; Grosseteste went further in pointing out that propositions should also be subjected to the test of *falsification* by experience. Many statements which cannot be verified (e.g., statements beginning "All . . .") can be falsified (e.g., by the discovery of a single counterinstance, such as the Australian black swan in the philosophers' favorite example). It is often far more decisive to look for the single destructive fact than to pile up more instances of the apparently obvious generalization.

Although Grosseteste furnished few practical examples of the use of his ideas concerning scientific method he did write a book on the rainbow in which he suggested that the colored bow might be caused by the refraction of sunlight in a cloud, just as colors are formed when light shines through a carafe of water. The study of refraction had always had a strong empirical content since the time of Ptolemy, and for a good reason: the amount by which light is bent (say in passing from air to water), and the actual appearance of prismatic colors, can only be observed at all by making simple experiments. The starting point for all medieval optics was provided by Alhazen, who made many sound investigations of vision, and of the properties of lenses and mirrors. Grosseteste's interest in optics was continued by his pupil Roger Bacon (c. 1214–1292), who gave a nearly correct value for the (determinate) angle between the sun, the center of the rainbow, and the eye; and by the Pole Witelo, who clearly obtained from an erroneous "law" certain figures which purport to be the result of measurements! The high point of medieval optics was reached by Theodoric of Freiberg (d. 1311), who reported a number of experiments in support of his theory that the rainbow is caused by two refractions and one reflection at the surface of spherical raindrops, and of his explanation of colors (taken from Averroës) as varying mixtures of lightness and darkness. His treatment of these two questions endured, with some improvements, until 1637 and 1672, respectively.

Optics was by no means the only science affected by reasoning from experiments or experience. But empiricism remained, for the Middle Ages, always rather an ideal than an actuality. Of no one is this more true than of Roger Bacon, who wrote largely about experiments, and made large claims for the merits of experimental science, yet himself accomplished little. He was fascinated by the experiments on magnetism of his friend Pierre de Maricourt (Petrus Peregrinus) and similarly by the mysteries of alchemical operations, but he did nothing in either of these fields himself. Bacon, equally eager to see mathematics employed more commonly in natural philosophy, understood little of mathematics. In short, like his greater namesake Francis four centuries later Roger Bacon was a visionary—not least in his dreams of what might be accomplished through learned technology:

> Machines for navigation can be made without rowers. . . . Also cars can be made so that without animals they will move with unbelievable rapidity. . . . Also flying machines can be constructed so that a man sits in the midst of the machine revolving some engine by which artificial wings are made to beat the air like a flying bird. . . . Also a machine can easily be made by which one man can draw a thousand to himself by violence against their wills. . . . Also machines can be made for walking in the sea and rivers . . . bridges across rivers without piers . . . and unheard of engines.[3]

Relying on ancient fables, Bacon thought all these feats had been effected in antiquity; for him, as yet, technological command over nature was by no means clearly separated from magic.

Though the frustrated mechanical ingenuity of medieval Europe did manifest itself in impracticable designs for flying machines, windmill-propelled vehicles, perpetual motion, and so forth, it was more constructively employed in building water wheels and windmills, used to grind corn, to saw lumber, to operate fulling and stamping mills and to work the bellows and hammers of smelting furnaces. Thanks to hints derived (probably) from China, Europe first made full use of gunpowder, the magnetic compass, paper, and the printing press. By the fifteenth century Europe was technologically superior to all other regions of the globe, except in certain crafts like the making of porcelain; and this technological development was starting to transform society in a way that had never occurred in history before. All this owed little or nothing to scholars, for the fruits of empiricism and inventiveness among learned men were small.* Besides what has been said of optics (which led to the useful invention of spectacles at the end of the thirteenth century) the very short *Letter on the Magnet* of Pierre de Maricourt (1269) is worthy of memory, for he described for the first time, and accurately, all the simple properties of magnets including the basic law "Like poles repel, unlike poles attract." The study of magnetism was important because, as it had been largely neglected by the Greeks and in Islam, it forced Europe to develop its own initiative; moreover, it lent strength to the notion of "attraction" (a mysterious force impelling bodies together) which was to become crucial later.

The chemical arts, concerned with the manufacture of glass, pigments, gunpowder and other pyrotechnic compounds, medicaments, and a number of chemical substances such as sulfur, alum, vitriol, and so forth, were wholly empirical. A great deal was learned here, and especially in alchemy, from the Arabs. "Geber" (a fictitious figure, based on Jabir ibn Haiyan, who lived in the eighth century) was the supposititious great master of these arts, while many genuine writings by Islamic authors were thoroughly pored over in the West. The language of chemistry is redolent of its origins. The most important single process thus to reach Europe was distillation; with its aid the simple mineral acids were discovered and above all "spirit of wine" (alcohol), which had become a panacea by the end of the fifteenth century. In alchemy distillation was the central operation, along with solution, cohobation, sublimation, precipitation, digestion, and many another wearisome maneuver to extract the quintessence of all things. Alchemy certainly had a theoretical basis in Aristotle's theory of matter, but as no one knew how to apply this theory to effect the transmutation of metals, the art of alchemy was a chaos of every kind of laboratory trick, using every kind of material

Like orpiment, burnt bones and iron filing . . .
Armenian clay, borax and verdigris . . .
And other useless nonsense of the sort
Not worth a leek, needless to name them all,
Water in rubefaction, bullock's gall,
Arsenic, brimstone, sal ammoniac,
And herbs that I could mention by the sack. . . .[4]

* The mechanical clock (fourteenth century) may be an instance of scholarly invention, though clocks were certainly made by blacksmiths.

The alchemical concept was philosophically respectable, for there was no known reason why the "substance" that was in lead should not acquire the "qualities" that characterize gold; but sensible men were always doubtful of practical success in this operation, believing rather with Ben Jonson

> *That* Alchemie *is a pretty kind of game,*
> *Somewhat like Tricks o' the Cards, to cheat a man,*
> *With charming.*[5]

Alchemy was in no intellectual sense the ancestor of true chemistry, which originated in the seventeenth century. For many it was applied (or "natural," that is, nondemonic) magic; to others it was a kind of applied religion with the Philosopher's Stone as a substitute for the Holy Grail. Alchemy was shot through with mysticism and mystery-making, lacking clear ideas and denying itself clear descriptions. Even considering techniques alone it may be doubted whether true chemistry did not derive more from craftsmen (apothecaries, metalworkers, and assayers, for instance) than it did from the alchemists properly speaking.

As alchemy suggests, experimenting could mean playing with fire, indulging a dangerous desire for familiarity with esoteric or forbidden knowledge. The experimenter might be a man who mastered strange powers (of optics, magnetism, or pyrotechnics) which the ignorant feared; he might even be a Faustian figure, driven by some unholy desire for knowledge to meddle in devilry. He was not far from the witch and was close to the astrologer, whose weird symbols and calculations of doom cast a similar shadow of disrepute upon mathematics. Philosophers (whose names were nevertheless invoked to give cover to spurious alchemical and astrological writings) did well to stand clear of such dubious activities.

Yet empiricism did play a part in the downfall of the Aristotelian synthesis that the thirteenth century had created and further strengthened by treating every part of science upon Peripatetic principles. Medieval experimental and observational resources were by no means powerful enough to destroy this synthesis factually, piecemeal, although they could suggest (for instance) that Aristotle's account of the rainbow (as formed by *reflection*) was false. In any case, the strength of Aristotelian science did not lie in any point-by-point correspondence with observed reality; it lay rather in its unity, in its use in all contexts of the same ultimate terms of explanation, and its logical consistency throughout with certain apparently indubitable metaphysical principles. The medieval synthesis could not have been overthrown by empiricism alone without an onslaught upon the philosophy of which Aristotelian science might be described as an extension or expression. And as, through the work of Aquinas especially, philosophy had become firmly interwoven with theology—even the Mass could be described though not explained on Aristotelian terms, the consecrated elements being transformed in *substance* though not in *form*—the theologians' doctrines too would have to be called in question.

Yet by the end of the thirteenth century there were doubts. There were Christian Averroists who refused to be satisfied by the orthodox interpretations of Aristotle's meaning. More important, with wider reading it became clear that the ancient world had not succumbed to Aristotle without protest. Galen, whose authority in medicine was not less than Aristotle's in philosophy and physics, had differed irreconcilably from him on many points. In some pronouncements, indeed, Galen's words sounded far more true and fit for the Christian ear than did Aristotle's. Again, the father of astronomy, Ptolemy, had described a system of the world very unlike Aristotle's, whose spheres (as every competent fourteenth-century astronomer knew) could not effect the observed planetary motions, whatever properties the philosophers might assign to them. It was the philosophers, however, who began the undermining of Aristotelian physics at its roots in metaphysics. Plato had held that "universals" (what we might call classes of things) were real as well as the individuals embraced within them. Aristotle taught that they were concepts, having a real existence only in the mind. Moreover, he held that certain truths about the nature of things (what we might call, though Aristotle did not, laws of nature) were necessarily true: the denial of them would result in illogicality or absurdity. Thus, he reasoned, the universe must be a plenum because it is logically impossible that there be a vacuum in it (a view reechoed by Descartes later); and for an inanimate object to move unless propelled by some mover would be absurd. Both these views of Aristotle were rejected by one school of medieval philosophers whose grand figure was William of Ockham (d. c. 1350).

These "nominalists" (or sometimes "terminists") held that the designation of a "universal" is merely the giving of a name to a group of things having some common feature. The "universal" was not even a concept; only individual things really existed, and only of these could the mind form concepts. Since to name a thing is not to explain it, the "universal" could not serve as a principle of explanation, as it so often did in Aristotelian science. More generally, the strict nominalists shifted interest from the general to the particular, or, extending the idea further, from the generalization to the single brute facts. Though the philosophical debate concerning the nature of universals had begun as far back as the tenth century, it was not until the fourteenth that the implications for science of the nominalist position became evident, especially through the writings of Ockham. For if the "universal" is not even a concept it cannot be known; we can at best only generalize about things, or arrange them under named groups, by considering individual entities and instances, that is, by using the method of induction. A priori knowledge of generalizations is impossible. Accordingly, in his writings on logic Ockham paid great attention to the improvement of reasoning from induction, continuing the thought of Grosseteste and Duns Scotus (1265–1308) and anticipating, in some measure, the more plainly expressed doctrines of Francis Bacon. The rule, or "razor," with which Ockham's name is always linked is the principle of economy: explanations should always be the simplest possible. He was also a strong advocate of the unity of nature (a principle to which Newton often appealed), from which it follows that like effects require like causes.

Ockham's logic foreshadows later scientific methodology, but he was no scientist himself, nor did he ever cease to be an Aristotelian. Before Ockham, as well as afterwards, there were other philosophers who advanced more direct criticisms of Peripatetic science, probing its theory of motion. Aristotle's necessary principle that no nonliving thing can move itself, true of objects at rest, true of heavy carts and boats, is seen not to be true of things once put in motion. The boat glides on when the tow rope is slack, the wheel spins when the foot leaves the treadle, and (above all) an arrow flies through the air after it has left the bowstring. Aristotle had explained all such discrepancies by making the medium (air or water) the mover: the original mover—bowstring, hand, or horse—in some way caused the medium to continue to move the body. He was unable to explain satisfactorily how this occurred; there was always a weakness here in Aristotelian physics. Fourteenth-century philosophers found in the Greek commentator Philoponos (fifth century) suggestions for an alternative theory, holding that a body in motion could continue without a mover, thanks to some power (*impetus*, or impressed force) contained within the moving thing itself. They recited the empirical reasons for doubting Aristotle's account, such as the greater distance traversed by a heavy projectile than by a light one like a feather, though only the latter could be blown by the air, and the fact that the medium, far from assisting motions, impedes them (as Aristotle himself had also recognized). They drove home the point that this peculiar, Aristotelian function of the medium involved a contradiction with its normal properties. Thus, impelled by experience and logic, they abandoned one of Aristotle's a priori principles of motion; nor was it all that they abandoned.

The two chief figures in this criticism of Peripatetic physics were both French, Jean Buridan (d. c. 1360) and Nicole Oresme (c. 1323–1382). Both taught at Paris, and both can justifiably be considered as precursors of modern physical science. In their hands the impetus theory of continued motion overwhelmed Aristotle's account, and they even extended it to the motions of the heavenly spheres. They argued that the Prime Mover of the universe, the ninth and outermost sphere, which Aristotle's theory required to drive the eight spheres bearing the observable heavenly bodies, was redundant, for if these spheres had been set rolling at the Creation and encountered no resistance, their impetus would cause them to roll forever. In this as in other contexts Buridan maintained that impetus was destroyed only by the resistance and friction which moving bodies ordinarily encounter. Others, however, supposed that impetus was spontaneously dissipated, like heat (with which it was often compared). It was generally recognized that the impetus of a given body was proportional both to its initial speed and to its weight.

One further application of the impetus theory was of great importance. Aristotelian physics had never stated a clear reason for the increase in a body's speed as it falls, though the fact was well known. Here was another case of motion without an apparent mover, a case assigned by Aristotle to the class of "natural" motions caused by the nature of things to return freely to their proper places. Thus, intrinsically heavy things would naturally fall to their proper places at the center of the Universe (the Earth) and intrinsically light things would rise to their proper places in the realms of air or fire. (Here Aristotle seems to be saying that everything requires a mover unless it doesn't.*) But it was not clear why the constant nature of a body should be the cause of a varying effect (the acceleration of fall), since this seems to contravene

* A further contradiction: the revolution of the heavenly spheres was also natural (though circular, while that of the four elements was rectilinear) yet they required a Prime Mover.

the principle that every effect is proportional to its cause, every speed (as Aristotle thought) to the moving force applied. The impetus theory explained acceleration in a new, mechanical fashion; in the first brief instant of time, its nature caused a free, heavy body to move a little; at the end of that instant it has acquired an impetus that would cause it to continue moving through the second instant at the same speed, but as its nature also impels it to descend, it now falls faster than in the first instant, and so on. This was as close as the medieval treatment of motion approached to the statement that a constant force produces a constant acceleration.

Could such an accelerated motion be described mathematically, just as a uniform motion is easily so described? In this the Greeks had failed, though they had mathematized other branches of physics, such as statics, hydrostatics, and optics. Knowledge of some of this achievement passed to medieval Europe through Islam, in such works as the pseudo-Aristotelian *Mechanical Problems,* "Euclid's" *Book of the Light and the Heavy,* and (after 1269) Archimedes' *On Floating Bodies.* The first Latin work on statics, by Jordanus de Nemore (early thirteenth century) already justified the law of the lever by a "principle of virtual velocities" and was able to state correct propositions about bodies standing upon inclined planes. Similarly, the Ptolemy–Alhazen geometrical optics was well understood in the West. But the treatment of bodies in nonuniform motion was a new venture that required the mathematical treatment of a *uniformly varying* quantity.

Foremost in this venture was a group of mathematically minded philosophers all associated with Merton College, Oxford, about the middle of the fourteenth century: especially Thomas Bradwardine, Richard Swineshead, Richard Heytesbury, and John Dumbleton. They were the first to distinguish between the causes of motion (dynamics) and the description of motion (kinematics). They carefully examined the concept of uniform acceleration, and they replaced Aristotle's law of motion $V \propto F/R$ by a new law, $V \propto \log (F/R)$.* Moreover, they developed and extended to motion a "calculus of forms" whose core is the "Merton Rule" that any uniformly varying form or velocity is equivalent to a uniform form or

* $V =$ velocity, $F =$ force, $R =$ resistance to motion. These formulations will not be found in the originals, of course, but they are justified by the more diffuse language found there. Bradwardine's new law gives motion only when $F > R$, whereas Aristotle's gave a finite velocity when $F = R$, which seemed absurd.

velocity, having the value of the midpoint of the varying form. A "form" in Aristotle's theory being an attribute of matter, like heat or color, the object of the calculus was the comparison of a form whose intensity changes with its extension with one whose intensity is constant. The calculus could be arithmetical or (as with Nicole Oresme) geometrical, which is easier to follow. Suppose as in Figure 4 that *AB* represents

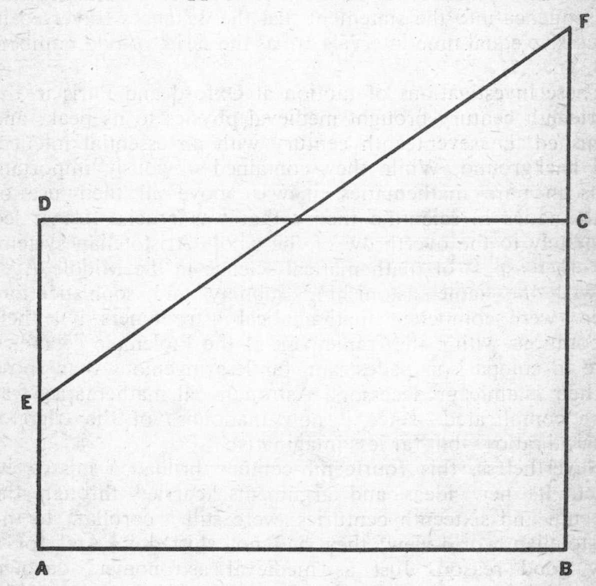

Fig. 4. The Merton Rule.

the extension of the form, then *EF* represents its changing intensity. The first discovery of the calculus was that *EF* could be defined as a straight line when this change is uniform (linear); if nonuniform *EF* would be a curve. Further, in the case of uniform variation the area *ABFE* is equal to that of the rectangle *ABCD* when *AD* is the mean between *AE* and *BF*; hence, by arguing that the effect of the form is the product of its intensity and extension, it is shown that the effect of the form varying from *AE* to *BF* is the same as that of the constant form *AD*.

Oresme went even further, teaching that if the form be velocity, then the extension (or first variable) against which its intensity is plotted must be time; and that the "effect" in this case is the distance traversed. According to Oresme's view, then, if the velocity increased from zero to V in time T, the distance traversed would be $1/2\ VT$. From their Rule the Merton scholars had noted that in the second time interval of such accelerated motion a body would traverse three times the distance passed over in the first; Oresme generalized this consequence into the statement that the distances traversed in successive equal time intervals are as the series of odd numbers 1, 3, 5. . . .*

These investigations of motion at Oxford and Paris in the fourteenth century brought medieval physics to its peak, and furnished the seventeenth century with an essential intellectual background. While they contained obviously important ideas in pure mathematics, it was above all their use of mathematics in scientific theory that was crucial, for it led ultimately to the overthrow of the whole Aristotelian system. No other aspect of mathematical science in the Middle Ages showed the same astonishing subtlety and sophistication. There were competent mathematical astronomers but their elaborations within the framework of the Ptolemaic geometry were in comparison pedestrian, far less ingenious than those of their Islamic predecessors. Astronomical mathematics was more complicated—since it now made use of the trigonometrical ratios—but far less imaginative.

Nevertheless, this fourteenth-century brilliance missed its effect. Its new ideas and arguments, carried through the fifteenth and sixteenth centuries, were still a corollary to the Aristotelian world-view; they had not shifted it. And for a very good reason. Just as medieval astronomers devoted much attention to the abstract motions of the spheres without paying much regard to the real paths of the celestial bodies, and just as medieval logicians discussed empiricism and deduction in the abstract, so medieval mathematical philosophers, gifted though they were, had developed a calculus of abstractions. They did not claim that their kinematic theorems applied to the movements of actual bodies, realizing that in many cases they could not. They deduced, correctly, consequences that would follow from certain suppositions, but they did not discover whether these suppositions were physically valid or not. Thus the fourteenth century scarcely impaired the authority of Aristotle's views upon

* This is equivalent to saying that the distance traversed is proportional to the square of the time elapsed, i.e., $s = 1/2\ at^2$.

what actually happens in nature, any more than Ptolemaic astronomy had detracted from Aristotle's cosmology as a picture of what the heavens are really like.

Some attempt was made in the fifteenth century to render this new mathematical physics more than arbitrary, as by Giovanni Marliani (d. 1483), who attempted some experiments upon its theorems; alas, they were not all confirmed. In the next century Niccolo Tartaglia (1500–1557) applied the impetus concept to the motion of projectiles, but his attempted science of ballistics was, apparently, ignorant of medieval kinematics. At about the same time the Spanish philosopher Dominigo da Soto first declared that falling bodies are freely accelerated and that the "Merton Rule" describes their motion. Later the same theme was taken up by the Italian mathematician Giovanbattista Benedetti (1530–1590). With him Galileo's lifetime is reached, and the overthrow of Aristotelian physics.

The medieval world recaptured Greek science, learning far more than the Romans ever knew. It mastered and improved all its techniques, save in advanced geometry and biological observation. It had made Aristotle its instructor in logic, metaphysics, physics, and cosmology, and it could not have done better. It had, moreover, prepared new skills that would enable men to explore the world and nature more deeply and widely than before, while at the same time laying the foundations of a new intellectual approach to science. To do so much was no small achievement of three centuries. But a new impulse was to start a violent movement from scholasticism, accelerated by all that was best in the scholastic tradition.

III

Biological Knowledge Before the Microscope

Chapter 7

THE DESCRIPTION OF LIVING THINGS

No less surely than physical science, biological science begins with the Greeks, but the pattern of its inheritance was somewhat different. For though the knowledge and the literature were there, they attracted medieval translators and commentators far less than did Greek physical treatises. No new biological theory comparable to impetus physics or fourteenth-century optics was to appear in the medieval schools, and there was no comparable revolt from the authority of the Greek masters—at least not then. Biology experienced a curious time lag: the full weight of ancient authority was not felt until the sixteenth century; the late sixteenth and early seventeenth century began to challenge Aristotle in biology as he had been challenged far earlier in physics, and it was only after the mid-seventeenth century that European biology became consistently superior to Greek biology. Even then, theories about the nature of life and living matter were to remain strongly Aristotelian until the late eighteenth century.

Though the word "biology" comes from Greek, it was not used before 1800; nevertheless, the Greeks had come closer to the concept than one might at first imagine. For they speculated about all living things, equally and impartially, and they framed theories and attempted analyses which assumed the underlying unity of all living matter. Here, once again, they appear to have been pioneers.

The Egyptians (and to a lesser extent the Babylonians) had possessed surgical skill, and some crude medical theory; there is no evidence whatsoever that they were interested in the serious study of the habits of plants and animals, or that they speculated about their structure and function. The Greeks on the other hand possessed a desire to understand living matter as much as any other part of the cosmos, though they found it less revealing of the order and harmony of nature than astronomy or mechanics. (Looking at worms is certainly a less lofty occupation than contemplating the heavens, though in the long run it may be more useful.) About the earliest speculations we know very little—far less than about contemporary theories of cosmology, for example. Evidently the pre-Socratic philosophers speculated about the origin of life; evidently by the early fifth century B.C. there was some interest in anatomical knowledge; but the references are scanty and ambiguous. Greek biology (as distinct from medicine, whose early history is plainer) begins with Aristotle, who here, as in cosmology, provided an all-embracing system so satisfying that it was to endure virtually intact until the seventeenth century.

It is not surprising that so powerful a mind as Aristotle's should have been capable of handling biological problems; it is surprising that the approach should have been so satisfactory and the results so sophisticated. Here at last the Aristotelian method of organizing a scientific investigation found the science to which it was ideally suited. Here classification, description, analysis, the search for causation—even the theory of motion—all found a role to play which illuminated rather than obscured the actual phenomena of nature. And here too Aristotle developed an intense interest in the empirical investigation of nature which was the obvious model for much of the empirical investigation in the physical sciences by the later members of the Lyceum and its affiliated research center, the Museum at Alexandria.

Aristotle approached the world of living things as a scientist, not as a medical man; he was interested in man only as one form of animal life. His classification was sophisticated and objective. On the one hand, he divided the cosmos into *animate* and *inanimate* objects (literally, objects with and without souls); the animate beings he defined as self-moved, motion being imparted by the soul. The "vegetable" soul, characteristic of plants, Aristotle associated with growth and nutrition; the "animal" soul with movement from place to place. Animals must be more complex than plants, having two varieties of soul; man further possessed a rational, intellectual soul. (It is obvious that Aristotle's notion of soul was quite different from that of the Judaeo-Christian tradition, though the two could readily be combined.) Aristotle, distinguishing between plants and animals, yet knew that the line between them was fine indeed, as for example in the case of sea anemones, which are plantlike in their inability to move from place to place but animallike in other respects. Animals themselves Aristotle classified mainly by their method of reproduction, achieving thereby a threefold division which is very nearly that of mammals, vertebrates, and invertebrates, though his lowest forms he thought possessed the power of spontaneous generation (these were the forms with no obvious egg-laying powers); he also spoke of "red-blooded" and "bloodless" (i.e., nonred-blooded) creatures. By emphasizing methods of generation (on which he wrote a complete treatise) Aristotle was able to recognize that whales, seals, and dolphins are not fish, but mammals; he also described and discussed the peculiar methods of reproduction of certain dogfish and sharks, as well as the breeding habits of cephalopods.

The basis of Aristotle's biological analysis is his *History of Animals,* a book whose title is more understandable when it is remembered that the word *history* derives from the Greek word which means "investigation." The *History of Animals* is a fascinating work, which deals with the classification of animals, their parts, internal and external, their methods of reproduction, and finally their habits, all displaying a tremendously wide knowledge. Clearly Aristotle must have drawn on others, though, in contrast to his work on physics, he names few sources and these mostly medical men. Either little had indeed been written on these subjects, or, more probably, what had been written was too factual and too little theoretical to merit discussion; there is scorn in his "some say," "others say," which precede tales and statements that are clearly false. Presumably Aristotle learned much of the habits of fish from fishermen, of domestic animals from farmers; but this could have been at best only empirical knowledge. It is clear from his descriptions that he dissected animals frequently, and occasionally he refers in the text to accompanying diagrams, now of course lost.

It is difficult to tell whether to admire most Aristotle's description of the bizarre habits of certain uncommon creatures—like the fishing frog which dangles its own tentacles as bait for small fish—which are quite correct, or the painstaking account of the habits of commoner creatures like bees. Aristotle knew of the distinction between drones and workers, and of the presence of a "ruler" (though he was not clear about the exact nature of the queen bee's functions) and described accurately the gathering and storing of honey, and the feeding of the young. In fact, Aristotle knew a surprising amount about insects, their life cycles and food. But the section which has always rightly attracted the most favorable comment is Aristotle's account of the development of the hen's egg. For he proceeded with such admirable care and such meticulous observation that it was really impossible to better his work until the advent of the microscope. He reported the first sign of alteration at the end of three days' incubation; at this time, the yolk had appeared to move to the end of the egg, and "the heart is visible like a red spot in the white of the egg. This spot palpitates and moves as though it were endowed with life."[1] He goes on to remark the development of the venous system, the appearance of the body, the changes in size of the eyes, the gradual appearance of individual organs and the position of the embryo, the vanishing of the yolk, and the final appearance of the bird. Nothing could be offered as a better example of descriptive zoology, even though the observed fact—that a red spot appears first—caused Aristotle to emphasize the heart as the central focus of the body, the first to live and the last to die, and led him to believe that it exercised far more control over the functions of the body than is in fact the case.

In the treatise *Parts of Animals* Aristotle does not describe the various organs of animals—that, as he notes, he had already done in the *History of Animals*—but rather, as he puts it, the causes of these parts. For this reason the treatise begins with a disquisition upon method in the natural sciences, a defense of the study of mortal beings (study of things divine—that is, the perfect, immortal heavens—being obviously superior) and a consideration of the theory of causation. Aristotle considered living things, like inanimate objects, to be a combination of form and matter, and therefore a complete explanation of them required consideration of both formal causes and material causes. The material causes are two: the matter out of which a thing is made, and the cause of the beginning of motion (which includes growth); the formal causes—especially important, as Aristotle believed, in living things—are the character of the creature (what makes it a worm, or a bird, or a fish) or of its parts, and the end for which it is made. This last, the final cause, was to Aristotle supreme. Though modern science has rejected teleological explanation, it cannot be denied that it is most useful when investigating the function of animal structure. Aristotle believed firmly that nature is purposeful, and that the harmony and beauty characteristic of the physical cosmos is found in living matter as well, and this he continually stressed.

So the *Parts of Animals* treats the purpose for which each organ or part is formed, and explains how the structure is related to the purpose. The little treatise *On the Motion of Animals* similarly deals with the cause of the particular kind of motion characteristic of each species of animal, and contains a preliminary disquisition on motion in general. Here is an attempt to relate animal motion to motion throughout the cosmos, as well as, at the other extreme, a briefly discussed distinction between voluntary and involuntary motions. In the *Progression of Animals* there is a more strictly biological examination of the ways in which animals move; this is not merely descriptive, but analyzes the way in which fish swim, birds and quadrupeds bend their legs, and makes a general comparison between the different classes, so that this can properly be said to deal with the mechanism of motion.

The biological concepts and method of Aristotle were followed by Theophrastos in his *Enquiry into Plants* (as the title suggests, it corresponds most closely to Aristotle's *History of Animals*)* and *Causes of Plants*. Theophrastos was Aristotle's successor as head of the Lyceum, and helped to preserve the tradition of scientific, as against purely philosophical, investigation. In his botanical works Theophrastos proceeded exactly as Aristotle had done, examining classification, parts, propagation, and natural history. Perhaps because of his subject matter, perhaps because of the changing times, Theophrastos was much interested in utility, but it is noticeable that medical applications of botany are discussed less than other matters. Highly characteristic are the discussions of the relation of flora to climate, a topic frequently dealt with in Greek medical and geographical literature.

The works of Aristotle on zoology and Theophrastos on botany mark the culmination of Greek biological thought in the fourth century B.C. Very strangely, the Hellenistic period

* In Greek the titles are very alike; the English title of the work of Theophrastos is a direct translation, because Theophrastos was little read before modern times; the title of Aristotle's work derives from the usual medieval Latin translation.

saw no further development comparable with that in the physical sciences or in medicine. Aristotle had shown that zoology was properly a part of natural philosophy, and had shown how to treat it as such; Theophrastos had proved the excellence and appropriateness of the Aristotelian method by using it in botany. They had no successors; Hellenistic investigation turned to man and neglected the rest of the living world. Later works fall into two categories, neither truly biological: lists of plants with their medical properties; and picturesque stories about animals.

Of these two classes of works, the first is the more truly relevant to science, or rather to applied science. Works of this kind appear as a systematization of traditional knowledge, built up over the course of centuries by herb- and root-gatherers. The earliest extant example is the treatise *On Medical Matters* (or *Materia Medica*) by Dioscorides, an army doctor of the first century A.D. The treatise of Dioscorides is simple, direct, and informative. He divides his subject into five sections, each section dealing with a group of plants. He names the plant, describes it, explains its medical uses, and provides an illustration to permit identification. The work is a handbook for the practicing physician or indeed anyone who wants to make use of natural products in the cure of disease; the illustrations made it particularly helpful, since there was, of course, as yet no standard or scientific system of nomenclature. Presumably the original illustrations were lovingly and carefully executed, and they were probably very accurate. The oldest illustrated manuscript known dates from the sixth century A.D.; five centuries' copying had not yet debased the illustrations, though it had somewhat formalized them, and the plants depicted are quite recognizable. This state of things, not surprisingly, soon changed. For the work of Dioscorides was immensely popular, and few copyists did more than copy —so that the illustrations became increasingly conventional. The Greek text was translated into Latin first in the sixth century, then again from Arabic in the eleventh century, and several more versions came into being during the Middle Ages; meanwhile the illustrations became stylized travesties. Soon, authors began to produce new versions, calling them *Herbals*, or *Gardens of Health*, each endeavoring to mine the wealth of Dioscorides as well as to add the plants familiar to northern Europeans.

The production of herbals was given tremendous impetus by the invention of printing. Now woodcuts could be used in place of hand-drawn illustrations; it was worthwhile to attempt an exact representation, since all copies would be alike and no errors need be caused by unskilled copyists. Writers sought out talented artists and tried to illustrate nature, not Dioscorides. The results, especially as the art of book illustration improved, were astonishing; there are few handsomer books than such sixteenth-century herbals as *The History of Plants* of the German botanist Leonhard Fuchs (1501–1566) or *The Living Portraits of Plants* of Otto Brunfels (1488–1534). Every country produced several such works at this time, all direct descendants of Dioscorides; the illustrations were usually superior to the text. Only a few men attempted to follow Theophrastos into botanical speculation, though his works, virtually unknown to the Middle Ages, were widely read in the sixteenth century.

It was actually his passionate, even reactionary, Aristotelianism that inspired Andreas Cesalpino (1519–1603), professor of pharmacology at Pisa. His botanical speculations were part of a systematic attempt to reform all science in favor of pure Aristotelian concepts. He saw Aristotle's doctrine of form and matter as the one necessary principle to an understanding of nature. Living matter he saw as a unity; for all of it contains a single, living principle, located at one spot in the animal organism. In higher animals this living principle is contained in the heart; in plants, Cesalpino located it in the "collar of the root," the point where stem and root join.

Plants and animals, Cesalpino thought, must be parallel organisms, a notion which led to some very strained analogies. But Cesalpino did classify his plants tacitly, though not overtly, by the kind of fruit produced, one of the first attempts to introduce categories based on a comparative study of forms. Other writers followed suit. Slowly, even medical herbalists began to introduce better-conceived systems of classification than that based on the alphabet; for example the Swiss Caspar Bauhin (1560–1624) tried to group plants by resemblances of external form. Slowly, botanists realized that plants required a complex series of classificatory subdivisions, and the binomial system, in which one name denoted an individual species and the other name a group, was introduced.

The first naturalist to write separate treatises on taxonomy was John Ray (1627–1705); he was not merely a botanist, but wrote on the classification of insects, birds, and fishes, as well as plants. With Ray begins the search for a "natural" system—a perfect method whereby all individuals sharing any characteristic are shown by their classification and nomenclature to be related. Ray particularly emphasized the distinction between monocotyledonous and dicotyledonous plants, but he did not rely wholly on the seed leaf, making use of root, flower, and fruit as well. Ray based his animal taxonomy on a careful consideration of comparative anatomy; he especially emphasized characteristic organs like feet and teeth. Though Ray did not believe that all living forms could be perfectly classified—for, like Aristotle, he thought all were connected in the scale of nature, so that sometimes one type would share characteristics with two dissimilar types, being midway between them—his system was in fact so much superior to all that had preceded him that it suggested the possibility of a truly perfect system. The most famous and influential attempt at a complete and all-embracing taxonomy, one that owed much to Ray, was the *System of Nature,* published in 1735 by the Swedish naturalist Carl Linnaeus (1707–1778). Here was a complete systematic taxonomy, "in which nature's three kingdoms are presented divided into classes, orders, genera, and species."[2] For Linnaeus, as for Ray, the species was the fundamental unit, reflecting God's original plan in the creation of the universe. Linnaeus laid great stress on the sexual characteristics of plants, reflecting thereby the work of Grew (1641–1712) and Camerarius (1665–1721); that is to say, he classed plants first by the number, proportion, and situation of the stamen. Further divisions were based on fruit, leaf, and so on. Linnaeus realized the immense importance of establishing a stable nomenclature, so that all botanists should have one and only one name for each plant; to this end he evolved elaborate and precise rules for nomenclature, which were almost universally adopted in his lifetime. His system was far from perfect, but it was accepted in broad outlines, and marks the culmination of the effort which had begun a century before to transcend the alphabetic lists of medical plants and to establish botany as an independent science.

Zoology in late antiquity pursued a different path from botany. Here, not utility, but entertainment was the cause of divergence from true scientific investigation. Even Aristotle had included many curious facts and stories in his *History,* though he tried to be critical and skeptical. Later generations were more credulous and more eager for entertainment. Aesop's *Fables* represent an older tradition than Aristotle; how the two traditions could be combined is well illustrated in the zoological portions of the *Natural History* of Pliny the Elder.

Pliny was a born teller of tales and delighted in good animal stories; since he wrote in Latin these quickly became part of medieval lore, and have been continually retold to the present day. In Pliny one may find the ostrich thinking itself invisible because its head is hidden; the bear licking its cubs into shape; the basilisk which kills with its breath; the lion which shows mercy to suppliants, and very many more. A fair sample is the treatment of the elephant:

> Amongst land animals the elephant is the largest and closest to man in intelligence; for it understands the language of its country, obeys commands, remembers duties, is pleased by affection and honors, and possesses besides what is rare even in man: honesty, wisdom, justice, as well as respect for the stars and reverence for the Sun and Moon. . . . They are also believed to understand the obligations of another's religion, since they refused to embark on board ship when going overseas until lured on by the mahout's sworn promise of return. And they have been seen when exhausted by suffering (even those vast frames are attacked by disease) to lie on their backs and throw grass up to the sky, as though deputing the earth to support their prayers. . . . The Indians employ the smaller breed, which they call the bastard elephants, for ploughing. . . .[3]

And so on, for a dozen pages. This is not zoology, though it is entertaining; Pliny's only guiding principle was a strong conviction of purpose in nature: every animal must serve some useful function.

This idea of use was soon adapted by Christian writers to moral and religious argument. Pliny's stories were well suited to allegorical treatment, and Christian animal stories abounded. The best-known collection of them is in a Greek work, probably of the second century A.D. Translated into Latin during the fifth century, it passed into medieval tradition under the title *Physiologus*. Here is the phoenix (known to Pliny, but without the allegory), which, when it is old, builds its own funeral pyre; out of the ashes comes a worm, which grows again into the phoenix: this symbolizes the resurrection of Christ. Here is the pelican, which first kills its rebellious young and then revives them with blood from its own breast, symbolizing God's anger at man for the Fall followed by His birth and death as Christ whereby He saved man with His blood. Here is not only the wonderful story of the unicorn, too strong to be captured by ordinary hunters, which docilely submits to a virgin, but the didactic reflection that the unicorn symbolizes Christ and the Virgin, of course, Mary.

Later writers seized eagerly on the possibilities inherent in combining wonderful stories of birds and beasts with a religious moral. Medieval bestiaries contain a gloriously credulous mixture of the fabulous, the miraculous, and the allegorical. Their authors had no critical standards, except that accounts must be entertaining: they knew the authority of tradition and believed that one could not go wrong in drawing on previous writers. When so much was unknown it was not at all difficult to believe that geese could be produced in the far north from barnacles or that the lion sleeps always with its eyes open.

By the thirteenth century, though bestiaries remained popular, certain steadying influences were at work to cause improvements. Aristotle's zoological writings were translated at the same time as his physical treatises; though the former works were less popular than the latter (there was no place for them in the medieval curriculum) they were read, and commented upon. The commentary of Albert the Great was extremely thorough; though he did not perceive that Aristotle wrote of Mediterranean species only, he did add brief accounts of some familiar northern species to Aristotle's *History*. Indeed, the best descriptions of the habits of animals came not from those writing systematic accounts, but from those who, mainly concerned with some other topic, happened to describe familiar birds and beasts. Thirteenth-century medieval travelers, going either to the far north or the far east (Marco Polo was the most colorful, but by no means the only man to describe the Mongols and their surroundings), met strange plants and animals unknown to the ancients and so were forced to describe them as they appeared. (Similarly, medieval illuminators could draw accurate representations of plants and flowers, but these naturalistic pictures had no connection with the illustrations to herbals.) One very striking example of observation and description is the handsome treatise *On the Art of Hunting with Birds* by Frederick II (1194–1250), Emperor of Germany and ruler of Sicily and South Italy with its brilliant trilingual culture. Frederick drew upon Aristotle where he could, and upon Arabic writers on falconry, but he wrote much from his own observation. He had a scorn of bookish men, and gladly criticized even Aristotle on the habits of birds; he also had a healthy skepticism of marvelous stories, and rejected the tale of the barnacle goose, commenting dryly that he had tried keeping logs incrusted with barnacles, but no geese ever appeared. He was, of course, primarily interested in falconry as a sport, and only secondarily in anatomy and zoology. Frederick was peculiarly gifted in this, as in other ways, and by no means represents the usual approach.

The introduction of printing meant that bestiaries could be illustrated and made more widely available, but a wider audience by no means meant a higher standard. Humanism, on the other hand, returning to the ancients, effected a revolution. First came new translations and editions of Aristotle's zoological works (rightly regarded as scientifically superior to his physical writings), of Pliny (criticized in the light of Aristotle and general knowledge), of Dioscorides, and, for the first time, of Theophrastos. Aristotle and Theophrastos showed what could be done, and sixteenth-century zoologists and botanists, more traveled and more critical than their medieval predecessors, recognized that the Greek authorities had described Mediterranean species, and that those in the north were different. This was a tremendous stimulus to the production of the zoological encyclopedia—like the *History of Animals* which Conrad Gesner published in the 1550s. This listed alphabetically every creature known to Gesner from either books or nature, gave its name in all the languages he knew, its habitat, appearance, physiology, diseases, habits, utility, diet, and any curious stories about it—all with authorities, who might be either ancient or contemporary, for Gesner, like a modern editor of an encyclopedia, commissioned naturalists to write special accounts. Gesner's information was, on the whole, good, and he was reasonably critical. Yet though he dismissed freaks of nature, Tritons and Sirens, he could not forbear including the sea-man and the bishop-fish whose head is shaped like a miter. Gesner's inspiration was twofold: he was a scholarly bibliophile and polymath, but he was also a man who delighted in nature, who wrote a book on the pleasures of mountain climbing, and who shared to the full the humanist belief that physical nature was made for man's enjoyment. An even larger work was that of Ulissi Aldrovandi (1522–1605), professor of pharmacology, first director of the natural-history museum and of the botanic garden at the University of Bologna. Aldrovandi labored long and lovingly at his work, himself writing volumes on birds and insects; the remainder of the fourteen volumes in which his labors are recorded were published after his death by his pupils.

Aldrovandi was the last of the encyclopedic naturalists; even in his lifetime the best zoological work was done by men who specialized. So Guillaume Rondelet (1507–1566), professor of medicine at Montpellier, concentrated on marine animals. His work is the outcome of a desire to vindicate Aristotle, which he was able to do on many points because he had access to the same Mediterranean species which Aristotle had initially described. His reliability was variable: he provided the first illustration of a dissected invertebrate—the sea urchin—while at the same time depicting the bishop-fish. Rondelet's younger contemporary, Pierre Belon (1517–1564), also a physician, traveled widely in the eastern Mediterranean and made a thorough survey of the flora and fauna he found there. In his treatises on fish he continued Rondelet's work, emphasizing the difference between animals known to Aristotle and those of northern waters; his illustrations are of a high standard of accuracy. Belon's work on birds is notable for its study of skeletal structure and its discussion of the homologies between human and avian skeletons.

Gradually the emphasis shifted until zoologists were no longer interested in rediscovering what Aristotle knew, but wished to explore the living world afresh. Accounts of the habits of animals were for popular works; serious zoologists turned to anatomical investigation. Examples are Ruini's beautifully illustrated *On the Anatomy and Diseases of the Horse* (1598), and the descriptions of animals and plants recently discovered in the New World; these culminate in Edward Tyson's series of monographs on the chimpanzee, the porpoise, the rattlesnake, the opossum, and others in the late seventeenth century. By the mid-seventeenth century the introduction of the microscope had essentially changed the character of zoological description. This was now concerned with the fine structure of organs, having passed beyond the stage of concern with mere gross anatomy, though that was always included, since much remained to be discovered.

One special problem fascinated zoologists in the late sixteenth and early seventeenth century—the problem of generation. Here once again they took their point of departure from Aristotle; consequently the emphasis was on animal rather than human reproduction. (Strictly speaking, both were discussed, but not always by the same people; few medical writers were prepared to undertake extensive studies of comparative anatomy. Consequently the best work on human generation came as a by-product of zoological investigation.) This work on generation falls into two parts: embryological and theoretical. The most interesting is the strictly embryological, especially if one includes the parallel work undertaken on the development of the mammalian fetus. In both birds and animals it was possible to pursue the method described by Aristotle, and follow the development of the embryo (in egg or womb) day by day throughout the period from conception to birth. So Fabricius of Aquapendente (1537–1619) undertook a wonderfully extensive series of investigations, writing on the development of the chick in the egg and the development of the fetus. He was not especially original, nor did he intend to be. His aim was to follow Aristotle step by step, extending where possible (as in his consideration of insect eggs), explaining, discussing, and interpreting. Probably his greatest contribution was his careful insistence on illustration: every step of the chick's development as far as he could see it from the third day onwards was carefully recorded in both words and drawing, and much the same was attempted for the mammalian fetus.

The theoretical problems Fabricius solved less well. Aristotle had thought that the chick is nourished by the yolk, though "made from" the white, a point which was much debated; Fabricius took the view that both nourished and neither made. A reasonable conclusion; except that it led him to decide that yolk and white are "of the same nature," and further to conclude that the material from which the chick was made—its starting point, so to speak—was the chalazae, the spiral bands running from the yolk to the lining membrane. Fabricius was often wrong, but he did grapple with relevant questions.

There was much further work—mainly on the mammalian fetus—between the publication of the works of Fabricius early in the seventeenth century and the publication by his pupil William Harvey (1578–1657) of *On the Generation of Animals* in 1651, but no further discoveries had been made: these were to await the use of the microscope. (Harvey used only a simple magnifying glass; microscopes were not much used before 1660, and much of Harvey's work was done before 1640.) Harvey was most interested in the development of the mammalian fetus; as physician to Charles I he made a study of deer from the royal parks, especially those killed in the hunt. This was an unfortunate choice; for he was unable to detect any change in the uterus for what appeared to be several months after fertilization; this not unnaturally made him doubt the necessity of renewed fertilization for each conception.* When it came to the hen's egg, however, Harvey's meticulousness in observation was better repaid. Most important, perhaps, was his realization that the chick embryo grew from the cicatricle in the yolk, thus solving the old problem, and he carefully watched its development. His account is more complete, and more accurate, than that of any of his predecessors; it gained peculiar importance from the fact that he regarded the development of the chick embryo as typical. For he believed that all creatures developed from an egg (though his notion of an egg was fairly elastic); this is probably less important as a blow against spontaneous generation, which was to have a long life yet, than it was as a blow for uniformity in nature. In spite of difficulties of evidence, he held that mammalian development was similar to that of the chick, and this assisted him to various improved theories about fetal life and development. Once again, further advances were to result from the use of the microscope in the years after Harvey's death; the final determination of the role of the egg in mammalian development had to await the nineteenth century.

* Such difficulties show why the experimental method was not always a reliable guide, and how difficult it was for seventeenth-century scientists to decide when to rely on their senses, and when on reason. Harvey would have done better had he had sheep rather than deer as experimental animals.

Chapter 8

THE FABRIC OF THE HUMAN BODY

The desire to combat human ills is as old as human history. Primitive man tried to cure disease with magical incantations. Later he looked for something material in nature—plants, or parts of animals, or mineral substances—which should assist magic. At first his choice of drugs depended on the logic of magic: a fancied resemblance to a human organ, the possession of some desired characteristic—these were enough to suggest that stone or plant might aid in reducing pain or swelling. There was also the view that disease was caused by evil spirits which could be expelled by the ingestion of nasty and unpleasant substances: this is the origin of many of the medicines based on dirt and excrements. Practice and custom suggested that some remedies worked better than others; the fact that a particular drug remained long in the pharmacopoeia might mean that (like foxglove, the source of digitalis) it had some genuine therapeutic value, though it might merely be a testament to the power of suggestion. Whatever theory of nature underlay this kind of medical treatment, it can hardly be said to be a rational theory.

Surgery, until quite recent times, was regarded as a distinct (though naturally related) activity. There is no theoretical basis to surgery: broken bones and wounds have an obviously mechanical cause, and an equally mechanical cure. This called for a craftsman to repair the damage. Certain cutaneous afflictions, like boils, also fell to the province of the surgeon. Gradually the surgeon learned to apply his knife, and his skill learned in reducing sprains and dislocations of joints, to cataracts, hernia and (in the Middle Ages) lithotomy. The surgeon's craft was regarded as distinct from the physician's art throughout antiquity and the Middle Ages; in fact, though the surgeon grew more learned as the centuries went by, the two branches, medicine and surgery, remained separate until the nineteenth century, and the surgeon's activities related to medicine only when, in the sixteenth century, he became accidentally involved in the treatment of a few specific diseases.

Greek medicine initially differed very little from that of other primitive peoples, with a simple religious interpretation of disease and its cure. Just as the Egyptian surgeon Imhotep became a god, so did the Greek physician Asclepios; by the sixth century B.C. there were numerous temples devoted to his worship and a complex cult associated with his name. Here the lame, halt, and blind came for cure as a religious exercise; the practice was that the suppliant (for he was that rather than a patient) underwent religious purification, prayed earnestly for assistance, and finally slept in the temple; if the preliminary rites were successful the suppliant dreamed of a cure by the god, and either awoke well, or awoke with the recollection of certain steps to be undertaken for a cure. The temples of Asclepios were always well supplied with votive offerings of limbs and organs in which the god had effected cures, as well as votive tablets commemorating them.* The temples of Asclepios continued to flourish throughout classical times and were the resort of hopeless cases, of hysterics, and of all with whom medicine proper was incompetent to deal. Rational historians of fifty or seventy-five years ago surmised that the priests were really highly trained physicians and surgeons, practicing stealthily by night in the guise of a god. Modern analysis of the evidence refutes this. We are now well aware of the possibility of psychological cures in very many cases; the situation was aggravated among the Greeks by the fact that physicians could do little indeed to *cure* disease; at best they could only alleviate suffering.

In considering Greek "scientific" (as distinct from religious) medicine, it is well to remember that medical theory and

* One of the latter gives an illuminating sidelight on temple practices. After the name of the suppliant comes the inscription "the god healed him of his blindness, but when he refused to pay the temple priests, the god struck him blind again. He returned to sleep in the temple, and paid his offering and was cured."[1]

medical practice must be judged by different standards. It is unfortunately true that the best and most ingenious of theories may lead to disastrous practice, and on the contrary a foolish and erroneous theory may conceivably produce quite satisfactory practice. Actual treatment of disease was, as the fifth-century physician Hippocrates called it, an art, and not a science; until the nineteenth century the successful physician was judged more by his relationship to his patient than by any other standard. Medical theory, on the other hand, bears an appreciable relationship to science, being effectively the zoology of man. Only medical theory will, therefore, be considered here.

The early history of the Greek physician is slightly obscure. We know that he too fostered a cult of Asclepios, but in this case as the founder of a brotherhood or guild of physicians, rather than of a priestly sect. The so-called Hippocratic Oath (which dates from about 500 B.C.) records the terms of indenture and the ethical code of the physician. It clearly applied to secular physicians who visited the sick in their homes, not to priests resident in a temple. It also specifically mentions one mode of therapy upon which Greek physicians always laid stress: diet. The Greeks (living in a region where agriculture was never rich) were perforce a frugal people, and inclined to assume that many of the discomforts associated with disease were the same as those associated with digestive disorders; consequently they expected a great deal from a proper dietary regime.

The earliest indications of a theory of disease—as well as the earliest indications of an interest in anatomy—date from about 500 B.C., and are associated with the Pythagoreans. As one might expect from the Pythagorean connection, the emphasis was on harmony: health is a state of balance, illness of imbalance. It was assumed that the bodily fluids were made up of what came to be called *humors*; good humor was a state of balance, bad humor of imbalance. By the time of Empedocles the humors were four: phlegm, blood, black bile, and yellow bile, associated with the four "temperaments," phlegmatic, sanguine, melancholic, and choleric; the survival of these adjectives preserves the ancient theory. These humors, being material, were further associated with the "qualities," moist, dry, hot, and cold, just like the four elements of matter.* Since disease was the result of an excess of some humor, ob-

* In fact, different writers associated the humors with different qualities; phlegm was always cold and wet, but the others were more variable.

viously it was desirable to deplete it, either by diet or by direct purgation of the humor involved. The latter was taken to be nature's way—as in the nasal discharge of colds, the breaking of boils and abscesses with discharge of matter, spontaneous nosebleeds, and so on. Many physicians felt that one should only assist nature with diet, warm or cold baths, and rest, others that one should assist nature by forcefully promoting the evacuation of humors. The former course was more favored in antiquity, the latter thereafter. The stupendous vomitings, purges, and bloodlettings of early modern medicine (c. 1500–1850) were a direct result of the humoral doctrine.

Another aspect of Greek medical theory was the concept of the *crisis,* which occurred on *critical days.* This may be partly a heritage of Pythagorean number mysticism; it is also associated with the prevalence of malaria, in which the fever fits recur every three or four days; and is primarily, of course, associated with the frequently sudden cessation of high fevers in favorable cases. Greek physicians laid great emphasis on the importance of recognizing the crisis, for they believed it possible to say certainly at that moment whether the patient would live or die. And to make prognosis easier, they kept elaborate records of case histories, which show meticulous care in observation, and that in the fifth century B.C. a fairly large number of disease entities was recognized.

Our knowledge of fifth-century medicine comes largely from a huge collection known as the Hippocratic Corpus. Because of the guild system of medical education, distinguished physicians had many disciples whose work was often associated with theirs. This is the origin of the Coan or Hippocratic school. Hippocrates (c. 460–c. 380 B.C.) was born on the island of Cos and after extensive travels returned there to teach and practice. Cos remained a medical center, and all the writings produced there—some, like the oath, pre-Hippocratic, many post-Hippocratic—became associated with the name of Hippocrates in late antiquity. It is now thought that Hippocrates himself was most probably the author of parts of the *Epidemics* (case histories), of *Airs, Waters and Places* (a treatise on climatology, intended as an aid to a physician to foretell the course of various diseases), of the *Regimen in Acute Diseases* (on diet and nursing), and of *On the Sacred Disease* (epilepsy). The last is particularly notable for its intense rationalism: Hippocrates insists that epilepsy is a natural disease like any other, with natural causes, and that it should, therefore, be treated like any other. There is the same common-sense rationalism running through all the treatises, together with an insistence on careful observation, objectivity, and a firm belief that nature is to be understood only through experience.

Other treatises in the Hippocratic Corpus—mostly belonging to the fourth century B.C.—deal with broken and (especially) dislocated bones, wounds and their treatment, and surgical instruments; there is a great deal of an aphoristic nature on medicine and man; and several further treatises on diet and its effect. The Hippocratic physicians and medical writers continued to lay stress on observation. At the same time there were other physicians more interested in medical theory, who tried to apply Aristotelian methods to the one science with which Aristotle had not dealt. It was perhaps a result of this Aristotelianism that the Alexandrian Museum (having initially a strong connection with Aristotle's Lyceum) became a center for research in human anatomy, now at last recognized as the fundamental science for both medical theory and medical practice.

The best work in human anatomy was done in the third century B.C.; later physicians looked back wistfully to the days of Herophilos (c. 290 B.C.) and Erasistratos (c. 275 B.C.), and described these men as the first to undertake human dissection. The Roman Celsus recorded that Herophilos was said to have practiced human vivisection; but since dissection of human cadavers was regarded as an unthinkable horror in Rome, and as Herophilos did perform vivisection experiments upon animals, it is most probable that the statement of Celsus tells us more about Roman opinion than it does about the historical facts. Certainly Herophilos dissected extensively; he had a particular interest in the relations of various organs, which led him to distinguish between the nerves and blood vessels, and further to denominate the brain as the center of nervous activity (rather than the heart, as Aristotle had believed). His study of the anatomy of the brain is commemorated in the names of several of its parts. Erasistratos is said to have studied pathological anatomy extensively, trying to find the origin of disease in lesions within some particular organ. He went further than Herophilos, and distinguished three sorts of "vessels": nerves, veins, and arteries, which he found associated with all organs. The arteries were not, he thought, blood vessels (they are often found to be empty after death), but carriers of *pneuma,* and his elaboration of the role played by *pneuma* caused his followers to be named pneumatists. *Pneuma* is a confusing term. It was the normal Greek word for *air;* it also meant *breath;* by extension it meant *life;* and, because of the Latin word used to translate it, it is often taken to mean *spirit.* (Not spirit in the sense of a ghost, but in the sense of a refined form of matter; a relic of Greek medical doctrine survives in the expression "to be in good spirits".) Erasistratos believed that the air taken in by the lungs was conveyed throughout the body by the arteries, just as blood was by the veins; air and blood were, as he knew, equally necessary for life. Erroneous as his view was, it is the first example of an attempt to explain the physiology of the respiratory and circulatory systems in any detail.

For our knowledge of the later progress of Greek medicine we are mainly dependent upon Celsus, who wrote in the first century A.D., and Galen, who wrote in the second. Both agree that physicians had tended to divide themselves into various schools, according to their prevailing views: the Dogmatists (or theorists) insisted that one needed to understand the physiology of the body in order to understand disease; the Empiricists rejected all theory in favor of experience and claimed that the physician needed to study only the individual patient and his symptoms; the Atomists argued that one needed to understand only the material composition of the body and its nutriments; the Methodists believed that the physician should study diseases, not symptoms. Clearly each group believed in what is a necessary part of a physician's knowledge; the weakness of each lies in the fact that in each case the physician practiced only his chosen aspect of medicine, and neglected all others. Probably the patient would have been best off in the hands of an Empiricist, who invoked the authority of Hippocrates—at least if he followed the genuinely Hippocratic doctrine of allowing the healing power of nature to effect the cure while the physician assisted to a relatively minor extent.

Claudios Galen (138–201) was the greatest of all Greek physicians; he came from Pergamon, a rich city in Asia Minor, but spent most of his life in Rome. Here he had an extensive fashionable practice, was physician to the gladiators, gave public lectures on medicine, and contrived to write prolifically. Nearly one hundred books by Galen have survived (some only in Arabic), and he himself mentioned the titles of very many more. His range of subjects was as wide as that of the whole Hippocratic Corpus: he wrote about himself, and his attitude towards his life and work; on medical theories; on anatomy; on physiology; on drugs; on medical practice, including numerous fascinating case histories—the list is all-embracing. Galen was far from being the objective, self-effacing narrator of the Hippocratic school; he is a vivid individualist who quotes his opponents in order to show them up as fools, and who never hesitates to claim credit where he feels credit is due. At the same time he was a brilliant experimenter who was able to apply the knowledge thereby gained to an understanding, and often a cure, of a specific disability.

Galen was at his best when dealing with structure and function, which he believed to be intimately related. In his desire to understand this relation and its consequences he pursued the study of anatomy with fervor and ingenuity. One of his most striking works is *On Anatomical Procedures,* a textbook of anatomical dissection that is a model of its kind. Here Galen begins with the skeleton—"as poles to tents and walls to houses," he says, "so are bones to living creatures"[2] —and then proceeds to the muscles, nerves, veins, and arteries of limbs and head. After this Galen turns to the internal organs, which he classifies as alimentary and respiratory, and then goes on to the brain. A related treatise, *On the Use of the Parts,* discusses the structure and purpose of each organ, without the directions for dissection contained in *Anatomical Procedures.* Galen's anatomical knowledge was far superior to that of Herophilos and Erasistratos; indeed, he was never tired of deriding the latter for thinking that the arteries did not contain blood; Galen was also aware of the pulsation of the blood in the arteries. He had a firm grasp of the nervous system, and knew the difference between the various nerves along the spine: one of his case histories describes his triumphant discomfiture of rival doctors who, being anatomically and physiologically ignorant, could not cure paralyzed fingers, nor understand how Galen could differentiate between the motor nerves of the finger and the sensory nerves of the skin covering the finger. Galen was extremely skillful at demonstrating the function of organs in the living animal, as witness his demonstration of the irreversible direction of flow of urine from the kidney to the bladder, which incidentally helps elucidate the functional structure of the kidneys, bladder, and ureters; or of the digestion of food in the stomach and the peristaltic motion of the intestine; or of the connection between the recurrent laryngeal nerve and the voice. Nothing so sweeping, so competent, or so imaginative had appeared in medical literature before; it was to be almost fourteen centuries before Vesalius was to attempt the same feat.

Galen's achievement is the more remarkable because he labored under enormous difficulties, of which he was well aware. He knew, none better, the importance of learning anatomy and physiology at first hand. As he wistfully remarked, "At Alexandria this is very easy, since the physicians in that country accompany the instruction they give to their students with opportunities for personal inspection."[3] But in Rome it was impossible for a physician to obtain a human cadaver for dissection. Galen recounts the shifts to which he was driven: he had studied (and perhaps stolen) a skeleton washed from a grave by a flood, as he had the skeleton of a would-be robber killed by his intended victim and left unburied by the roadside. If such things failed, then Galen advised falling back on animal dissection, preferably of apes, choosing those most like man. This expedient he thought second best, only to be commended because it permitted a physician who had dissected apes to recognize human bones and organs by their similarity to animal organs. Galen was indefatigable in pointing out that animal dissection was at best a poor substitute for human dissection, and that the practicing physician should be quick to detect both the differences between human and animal structures, and those found among the various animals. Himself restricted to animal dissection, combined with observation of his patients, he was most accurate on bones, muscles, nerves, veins, and arteries, which are externally accessible; least accurate in regard to internal organs, which he could not verify in human subjects.

Galen's insistence on correlation between form and function and his firm belief in teleology were thoroughly Aristotelian in concept, and as with Aristotle, these principles served Galen well. He was certain that every organ of the body is ideally suited to its purpose and works perfectly (for Nature has designed an ideal structure in the body), and therefore every organ must reveal its function in its form. But there are occasions when the matter is more complex, when the most obvious aspect of an organ is not its most important, and when one's first impressions of form are not those which most closely reveal function. In these cases, Galen's ideas become somewhat uncertain and difficult to interpret; unfortunately, just such a case is his account of the cardiovascular system, which is central to an understanding of human physiology—though Galen did not entirely comprehend its importance.

In contrast to Aristotle, who laid complete stress upon the primacy of the heart, Galen began his reconstruction of physiology with the liver, for him the center of nutrition. Food when digested in the stomach and intestine produces chyle, which then proceeds through the portal vein to the liver (Fig-

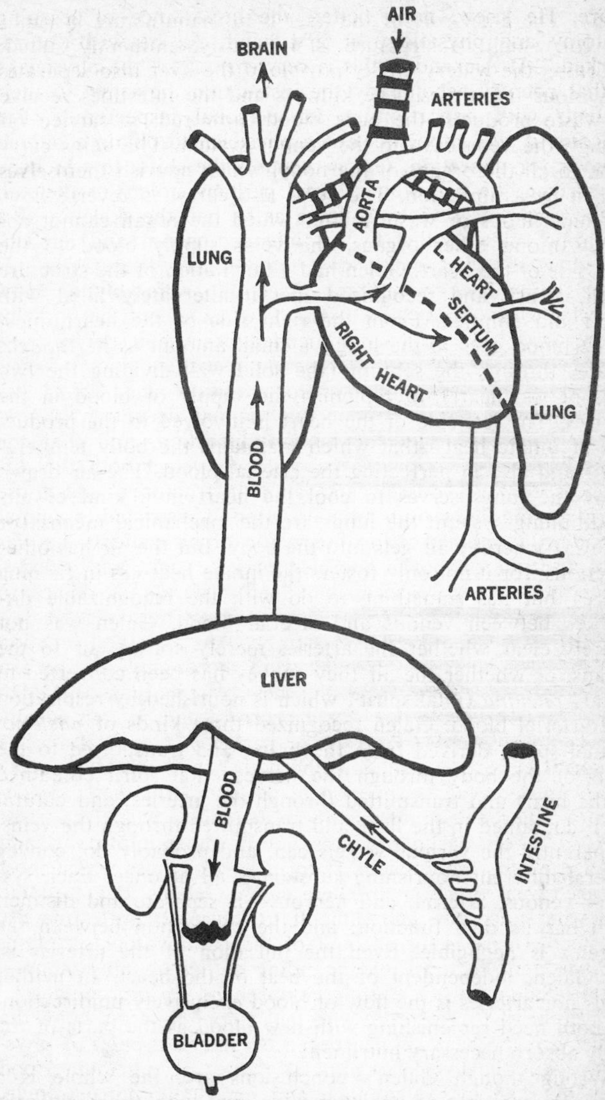

Fig. 5. The principal organs in Galen's physiology.

ure 5). The chief role of the liver is to manufacture blood, a task for which its structure is, obviously, eminently suited; in turning the watery chyle into blood the liver also separates off various impurities. The kidneys and the intestines receive the waste products; the pure blood, now red, is carried off through the vena cava to the venous system. The veins carry blood to all the organs of the body, which nourish themselves from it by assimilation. The veins also carry, in reverse flow, any superfluous or waste matter which the organ cannot absorb. Among other organs, the veins supply blood to the right side of the heart. Galen had a fair notion of the structure of the heart, and recognized that it alternately filled with blood and emptied. From the right side of the heart, most of the blood goes to the lungs; a small amount as he thought trickles through the septum (the solid wall dividing the two sides of the heart) to replenish the supply of blood in the arteries. The left side of the heart is involved in the production of innate heat—that which maintains the body temperature—and also in perfecting the arterial blood. The air drawn in by the lungs serves to cool the heart in a kind of air-conditioning system; the lungs are the mechanical means (or bellows) whereby air gets into the body. But the air has other functions; for it not only fosters the innate heat (as in fanning a fire) but has something to do with the recognizable difference between venous and arterial blood. Galen was not entirely clear whether the arteries merely convey air to the organs, or whether the air they convey has been converted to "vital" *pneuma* (vital spirit) which is nourished by respiration and arterial blood. Galen recognized three kinds of *pneuma:* animal spirit, derived from the brain, and transported to the parts of the body through the nerves; vital spirit contained in the heart and transmitted through the arteries; and natural spirit contained in the liver and transmitted through the veins. Apparently the various vessels can, and probably do, convey several different nourishing substances all at once. Each system—venous, arterial, and nervous—is separate and distinct; each has its own function; and the connection between the systems is negligible. Even the pulsation of the arteries is, for Galen, independent of the beat of the heart. In neither veins nor arteries is the flow of blood exclusively unidirectional; both need replenishing with new blood as the parts of the body absorb necessary nutriment.

Wrong though Galen's conclusions are, the whole is a powerful example of an attempt to deal with the body as a single harmonious organism, with each part contributing its share, and each dependent upon the other. The treatise in which these ideas are chiefly discussed, the *Natural Faculties,* is a blend of observation, experiment, and reasoning from principles in a characteristic and fascinating combination. The observation and experiment are so wholly admirable, the criticism of other men's views so just, that it is difficult not to find the whole argument utterly convincing. It is no wonder that centuries later the works of Galen, like those of Aristotle and Ptolemy, were to provide a fertile field for the commentator.

The spirit of inquiry vanished from medicine even more abruptly than from the physical sciences. Interest in medical practice survived: hence the popularity of Dioscorides, who provided a handy compendium of drugs. Celsus survived in Latin (though he was in fact little read), and there were a number of Latin translations of minor dietetic, prognostic, and therapeutic works of Hippocrates and Galen available for the encyclopedists to quarry. In the East, Byzantine medical writers like Alexander of Tralles (sixth century) and Paul of Aegina (seventh century) codified an educational tradition by combining observations from their own experience with those of Hippocrates and Galen in handbooks of medical theory and practice; the practice was often sensible, and new drugs were introduced, yet the theory became ever more drastically simplified until anatomical and physiological investigation all but disappeared.

The Hippocratic Corpus and the numerous works of Galen were nearly all translated into Arabic; indeed, several of Galen's books survive today only in Arabic, the Greek being lost. But Islamic medicine suffered from a grave handicap: dissection was forbidden, as was the pictorial representation of the human body, and, for the truly devout, the representation of animals. Islamic physicians, like their Byzantine contemporaries, turned to the treatment of disease, relying on the ancients for medical theory. A number of Islamic medical writers are notable for their descriptions of disease entities: thus the Persian Rhazes (865–925), a prolific writer on many subjects, described and differentiated smallpox and measles in his treatise on plague. The Muslims were keen and vigorous pharmacists, who introduced a wide variety of vegetable, animal, and mineral preparations into the pharmacopoeia, as well as preserving the use of the drugs common in antiquity. The most influential Islamic medical work was the *Canon* of Avicenna (980–1037), a great compendium;

although it contains a discussion of the nature of medicine, this is mainly devoted to definition, and the bulk of the work deals with diseases and their treatment, including the recognition and preparation of the necessary drugs. Avicenna's *Canon* remained the great standard work in Islamic countries until modern times; in Latin translation it was widely read in Europe and remained popular into the seventeenth century.

The medical tradition ran very thin in early medieval Europe. There were numerous books on herbs and drugs, often with a strong magical element, and a few discussions of medical practice, but nothing of note until the Muslim influence pervaded Europe. Medical works were among the earliest translations: Rhazes and Avicenna shared favor with Hippocratic and Galenic writings; indeed the Islamic tradition was dominant. The earliest works translated were all therapeutic and pharmaceutical. Very little indeed of Galen's anatomical and physiological work was available; only a very much abridged version of *The Use of the Parts* had been translated by the end of the thirteenth century. Medieval physicians were forced to reinvent the tradition of anatomical dissection for themselves; having done so, they turned, slowly, to Galen for assistance, so that translations of Galen's anatomical treatises were at last made.

Because medicine formed a part of the university curriculum, both medical education and the profession itself were organized along quite different lines from those prevailing in the ancient world. The professor was a "doctor," that is, a man who had completed first the university arts curriculum and then the advanced medical curriculum. Possession of the M.D. degree effectively licensed a man to practice as a physician (and improved the scale of his fees as well). The doctor was thus a learned man, and his long robe an outward sign of his education. The physician, at the top of his profession, made use of the services of lowlier men. There was the pharmacist, druggist, or apothecary, who prepared medicaments which the physician prescribed; there was the surgeon, who bandaged wounds and attended to fractures; and there was the barber-surgeon, whose chief activity was letting blood at the physicians's orders, though he also performed operations, like lithotomy, removal of cataracts, and correction of hernia. In the fashion of medieval society, all these professions were organized into guilds and their functions strictly delimited—though the physicians frequently had to battle to preserve their prerogative. There were also, of course, many unlicensed practitioners, especially in the villages—wise women and folk doctors whose prescriptions were simpler (and cheaper) than those of the physician, and often surprisingly similar.

Since the medical faculty was a part of the university, normal scholastic methods were used in instruction. The professor read a text, and commented upon it at length. The text might be by Galen or Hippocrates or Avicenna, or by any one of a dozen or more minor Greek and Arabic writers (but not Celsus); it dealt either with aphoristic considerations upon the nature of medicine, or with materia medica, or with the diagnosis and treatment of disease. Initially instruction was purely verbal, but gradually an element of practical demonstration, especially in anatomy, crept in. This appears to have begun at Salerno, which was a medical center as early as the late tenth century; here the first medical texts were probably translated from Arabic, and the names of numbers of famous, though rather shadowy, physicians and surgeons are associated with it. Here at least some animal dissection, especially of the pig, was carried out in the twelfth century. By that time, northern universities—Bologna, Padua, Montpellier, and Paris—were beginning to establish medical faculties, destined to be world-famous in the thirteenth and fourteenth centuries.

Human anatomy was at first investigated only in the course of postmortems. Initially, these were performed at the request of the faculties of law, investigating the death of important personages; in the fourteenth and fifteenth centuries postmortems were frequently performed at the instigation of the patient's family or even of the physician anxious to study an interesting disease; many of the medieval miniatures apparently showing dissection scenes in fact represent postmortems. But by 1300 human dissection was recognized as an essential ingredient of medical education, and all good medical faculties provided one or two public dissections which students were first allowed and then required to attend every winter. After 1316, there was available a most useful guide to dissection in the *Anatomy* of Mondino dei Luzzi (c. 1275–1326), who himself lectured at the University of Bologna. It is a technical treatise, based upon the fragmentary version of Galen's *On the Use of the Parts* which had been so imperfectly rendered into Latin that most of the names of organs were still Arabic. Mondino gives fairly good directions: he begins with the abdomen, then proceeds to the thorax and thence to the head; last of all come the limbs. He mentions only the more important organs, and the directions are not precise; this is an outline rather than a detailed handbook. But it served as an admirable text for professors to read and comment upon, while a dissector opened the cadaver and displayed the organs the professor mentioned. Much of Mondino's anatomy is animal, not human; but precision was not his aim, and no one noticed the difference. Indeed, when Galen's *On the Use of the Parts* was translated in its entirety directly from the Greek half a dozen years after Mondino's *Anatomy* was written, it was little heeded, probably because it seemed too complex and cumbersome. It was only slowly that it gained popularity; by the fifteenth century, however, it had become a standard text, though it did not replace Mondino's work, to which it was, indeed, supplementary.

The invention of printing did nothing at first to change the position. Mondino's *Anatomy* was printed; by the end of the century there began to appear commentaries on Mondino, and the first note of criticism is heard. Meanwhile, humanists were as busy on medical works as on other subjects. There were new translations of the previously known works of Hippocrates and Galen, now completely Latinized, the Arabic names for diseases and organs having been replaced by classical Latin terms thanks to the assistance of the newly rediscovered Celsus. *On the Use of the Parts,* in fresh translation, became very popular after 1500 and advanced medical lecturers encouraged their students to read not only this, but also such newly translated and previously unknown texts as Galen's *On the Motion of the Muscles* and *On the Natural Faculties*. The most important and influential "new" work was Galen's *On Anatomical Procedures,* first translated by a professor at Paris and published in 1531. The new light shed by Galen produced a marvelous revival of interest in anatomy. Just as fifteenth-century astronomers tried to master Ptolemy as a first step towards mastering the heavens, so sixteenth-century anatomists tried to master Galen as a first step towards understanding the fabric of the human body. This was excellent when it meant following Galen's precepts and studying the human body from skeleton (little studied before the sixteenth century) through muscles, veins, and arteries to alimentary, respiratory, and nervous systems. It was less good when it meant—as unfortunately it did to some—the intensive study of what Galen wrote without any attempt at repetition of his dissections. This was especially regrettable since human cadavers were now in fairly good supply (most medical schools had ample access to the bodies of criminals after execution), in marked contrast to Galen's Rome; experience soon showed that Galen, relying on animals, had made many errors, but those who clung to Galen as supreme authority were happier "seeing" what he described than admitting that Galen had erred. So Galen suffered the same fate as Aristotle, though in both cases some men, happily, followed the spirit rather than the letter of the Greek authority.

A separate source of inspiration arose from the demands of art; the new school of painting emphasized naturalism in representing the human form as in representing external nature. Artists were, naturally, most interested in surface anatomy, especially in the structure of the muscles. Only one important artist ever went far enough to rank as an independent investigator—Leonardo da Vinci (1452–1519), who turned, probably under the influence of Galen's *On the Use of the Parts,* to an attempt to understand the function of organs. Leonardo had the advantage of the artist's eye which is trained to observe, and he was a good anatomist, though not so exceptional as was once thought. What is extraordinary about Leonardo is the artistic merit of his notebook pages; what his completed work might have been like is impossible to say, for he published nothing and showed few people his work. Any influence he may have had was on anatomical illustration. Many artists in the sixteenth century developed this specialized skill, and anatomical books of the times are wonderful picture books. The illustrations range from the spirited but improbable plates illustrating muscles in *A Short but Very Clear and Fruitful Introduction to the Anatomy of the Human Body, Published by Request of his Students* (1522) by Berengario da Carpi (c. 1460–c. 1530) to the rich and appealing illustrations (probably by Jan Stephen van Calcar) of the great work of Vesalius *On the Fabric of the Human Body* (1543), better known by the abbreviation of its Latin title, *Fabrica.*

The *Fabrica* of Vesalius was not unique as a work on anatomy, nor as an illustrated work, nor as one which denounced numerous errors made by Galen, nor as one in which new discoveries were recorded. Berengario had done all these things, and he was one among many. Nevertheless the *Fabrica* is a landmark, if only because of its opulence. It was bigger, more complete, better illustrated, and contained more denunciations of Galen and more new discoveries than any previously. Vesalius was well read in Galen: he had studied at Paris with Guinther of Andernach who had first translated Galen's *Anatomical Procedures,* and he had edited one of Galen's treatises; and he admired him—so much, indeed, that he felt it was wronging Galen to follow his errors instead of following his method and correcting his errors. He also followed Galen in trying to relate structure and function, so that the *Fabrica* is at once a great anatomical handbook and a treatise on physiology. Vesalius was convinced (like Galen again) that the body is a harmonious whole, and every part is to be examined in relation to the whole. So the illustrations of bones proceed from complete skeleton to the smallest earbone; the illustrations of muscles show layer after layer; the whole visceral cavity is shown with everything in place before each organ is examined and described. Vesalius saw the whole body as a going mechanism, the venous system as analogous to a water supply, and this in turn led him to take a severely objective and rational view of all organs and their structure.

This is not to say that Vesalius was always right. Like Galen, like Leonardo, like all sixteenth-century anatomists, he often carried over what he observed in animal dissections into human anatomy. Like them, he was confused when it came to structures where Galen was right for animals but wrong for human beings. (So Vesalius always "saw"—and had the artist portray—the right kidney in man as lying higher than the left; this, as Galen had intimated, is "true for all animals," but not for man.) It is difficult enough to observe accurately without preconceptions; how much more difficult when one observes with the memory of what another has said. Vesalius did not find it easy to make innovations, ready to do so though he was; it is the more to his credit that he made so many, and was so often correct.

Good Galenist that he was, Vesalius made a thorough investigation of the structure and function of the various physiological systems, going more deeply than Galen into such questions as how the fiber structure of the veins is responsible for the transmission of blood through the venous system. The essential outlines of his physiology do not differ from those of Galen. But he raised some problems which were to be of subsequent significance: the most important was that he could not understand how any blood got from the right side of the heart (connected to the venous system) to the left side of the heart (connected to the arterial system). Galen had thought some blood trickled through the "pores" in the septum; where Galen described pores, Vesalius saw merely blind pits. Curiosity and thoroughness prompted him to explore the situation, for pores were physiologically necessary; but he found no evidence that these pits penetrated the septum. Nor did he find any other way in which the septum might be pervious. Yet some blood must get from veins to arteries . . . the matter was very curious. Another problem was caused for Vesalius by the possible theological implications of respiratory anatomy. Galen's *pneuma* or *spirit* was at this time often rendered as *soul;* this was especially the case with the "animal soul" elaborated within the brain, but there was some confusion over the connection between the "vital soul" and the respiratory functions of the heart. Vesalius boldly declared his right to discuss anatomy without regard to potential theological controversy; the result was thorough, but inoffensively orthodox.

Just how tricky the theological implications of the anatomy of respiration might be was amply demonstrated by a physician who was also a theologian, Michael Servetus (c. 1511–1553). Born a Spaniard, he lived most of his life in France; destined for the law, he turned instead to radical unitarian theology; university-trained, he earned his living by humanism, astrology, and medicine in unorthodox succession. The paradoxes are typical of his character. On only one point was he consistent: he would not cease from theological controversy, nor refrain from attacking all trinitarians, Catholic and Protestant alike. The book which brought him posthumous fame as a physiological innovator also brought him martyrdom in Calvin's Geneva; it was called *On the Restitution of Christianity* (1553) and (naturally) dealt with unitarian theology. This led Servetus, through logic perhaps only he could understand, into discussing first how the divine spirit is breathed into man by God, next how man breathes and acquires "spirit" in the blood. Educated in the medical school of Paris, by the same teacher as Vesalius, Servetus was both well trained in Galen and aware of the solidity of the septum. He concluded that the blood did not seep through the heart; instead it made a long journey from the right side of the heart through the lungs and so round to the left side of the heart. This statement of the pulmonary or lesser circulation (though in fact Servetus did not believe that all the blood so passed from veins to arteries; only a small portion) is the first printed account; like all other sixteenth-century anatomists Servetus triumphed in proclaiming a truth unknown to Galen.

Unfortunately, few could read Servetus; all but four copies

of his book perished with him, and not many anatomists were likely to be tempted to read a work of radical and heretical theological controversy, one in which, after all, no one could anticipate finding anything of anatomical interest. Fortunately the idea which had occurred to Servetus out of the logical need to find an alternative path to that through the septum, now ineluctably blocked by Vesalius' determined denial of pores, occurred to others for precisely the same reason. Ironically, the first to lecture on the pulmonary circulation, and the first after Servetus to publish the discovery, was Vesalius' successor at Padua, Realdus Columbus (1516–1559). In 1537 he had failed in open competition with Vesalius for the job which he only received when Vesalius left to be physician to the Emperor Charles V; it is no wonder that he enjoyed pointing out the mistakes of Vesalius as much as those of Galen, nor that his mind was receptive to new observations of things unperceived by Vesalius. Columbus was an excellent lecturer on anatomy, and as skilled in dissection as any of his contemporaries; he had a lively if polemical style, and a deep interest in anatomical structure. He had no doubt at all that the blood did not penetrate the septum, but that it went directly from the right side of the heart to the lungs and thence to the left side of the heart. With pardonable (if mistaken) pride he observed, "This fact no one has hitherto observed or recorded in writing, yet it may be most readily observed by anyone."[4] Columbus also took a simple view of the physiology involved: in the lungs, he thought, air was mixed with blood and sent to the left side of the heart, but the lungs, not the heart, are responsible for drawing off the vital spirits from the air.

The views of Columbus were published posthumously in 1559; many, like Cesalpino, had attended his lectures and learned of his views before publication. Cesalpino, devout Aristotelian as he was, felt that this view consorted better with the Aristotelian primacy of the heart than Galen's view: he therefore eagerly adopted it. Trying to make sense of Aristotle's somewhat vague description of supposed physiological changes in sleep, Cesalpino suggested that when we are asleep "a large supply" of blood is conveyed to the heart through arteries and veins; when we are awake an equal quantity is conveyed from the heart to the arteries and nerves. Though on the strength of this argument some have concluded that Cesalpino must have arrived at the concept of the circulation of the blood, in fact he did not; he thought of a tidal flow, with reversal on sleeping and waking. And he found no need to offer anatomical evidence in support of his supposition.

Vesalius had suggested that the venous system resembles a water supply. This very natural analogy was to offer a useful model to late sixteenth-century anatomists, though it was not until Harvey that anyone saw the exact way in which the analogy ought to be applied. The difficulties of correct interpretation of observed fact are excellently illustrated by the gap between the first observations of little membranes in the veins and the realization that these were valves which acted to ensure unidirectional flow. Numerous anatomists detected membranes in various veins; the most complete account of their structure, and the first reasonable interpretation of their function, was contained in a very short treatise *On the Valves in the Veins* published in 1603 by Fabricius of Aquapendente, complete with illustrations. Fabricius was so struck by these membranes that he first ascertained that all the veins possessed them and then tried to explain their purpose. He called them, not valves, but floodgates, and thought their function was to control the volume of the blood supply, so that the extremities should not become gorged with excess blood, just as a dam in a river controls the water above a water mill. Like all sixteenth-century anatomists Fabricius considered the heart to be part of the respiratory system, not of the venous system; hence he failed to connect the control of blood supply with the possible action of the heart.

It was because he believed (with Aristotle) that the heart was the center of the organism, and because he tried (with Galen) to harmonize the structure of the heart with its function, that William Harvey (1578–1657) was able, not only to interpret correctly the action of the venous "floodgates," but to go further and relate the heart to the blood supply. As an old man he remembered that he had begun his investigations by trying to understand the purpose of the venous membranes, being dissatisfied with the explanation of his teacher at Padua, Fabricius. An equally important motive was his desire to reinstate the heart as an important organ in its own right; for those before him had, he complained, regarded it as merely an appendage to the lungs. This made him examine the structure of the heart with fresh eyes, and in a new way; from structure, he quickly became aware of a function quite different from any advocated before; and having deduced this function, he proceeded to demonstrate it by anatomical investigations.

Harvey published his conclusions in 1628 in a work aptly entitled *An Anatomical Essay on the Motion of the Heart and Blood in Animals*—a title which by itself suggested novel concepts. He had first announced his ideas to fellow members of the College of Physicians ten years before. His lecture notes show that his argument ran like this: the structure of the heart makes it plain that blood goes from the right ventricle of the heart through the lungs and so to the left ventricle and aorta—direction of flow being governed by the valves in the heart and veins—and this continuously, not in small trickles. The venous membranes act as valves to preserve unidirectional flow from arteries to veins; this could be demonstrated by ligating an arm either loosely, when the veins swell below the bandage, or tightly, when the veins are depleted because the blood cannot get through the arteries to fill the veins. "Whence it follows," so Harvey triumphantly proclaimed, "that the movement of the blood is constantly in a circle, and is brought about by the beat of the heart."[5] This involves two novelties: the concept of circulation and the idea that it is the beat of the heart which pushes blood out into the arteries. (Previously it had been thought that the heart sucked in the blood required by the left ventricle and that the pulse reflected arterial action.) Harvey tested his conclusions by countless observations. He found it particularly useful to examine cold-blooded animals: their hearts would continue to beat even when removed from their body, and the fact that they breathed through gills, rather than lungs, helped in distinguishing between respiratory and cardiac functions. True to his belief that Galen was right to associate structure and function, Harvey showed that the heart was made of muscular material; it acted, in fact, only as a muscle, driving the blood about the body. Further, watching the heart drive the blood and considering its small size, it occurred to Harvey to make a simple calculation. Suppose (conservatively) that the human heart held two ounces, and beat eighty times a minute; in the course of a day the blood pumped by the heart would weigh many times more than the man whose heart it was. This was impossible: the liver could not continually produce so much blood, nor the body absorb it.

There remained nothing except to conclude that the blood circulated, taking its origin (effectively) from the left side of the heart, proceeding through the arteries to the veins to the right side of the heart, to the lungs, and thence back to the left side of the heart again. Harvey looked for, and found, endless confirmation; and the argument with which his book is filled is almost tiresomely convincing. Only one piece of evidence escaped him: he could not detect the connection between arteries and veins. He could only suppose that the arteries, growing ever smaller, ended in a "marshland," an area of spongy tissue where the blood lay free outside the blood vessels; and that it seeped into the smallest veins which then took up the task of conveying it back to the heart. Once again, biological investigation waited for the microscope; the capillaries were detected in 1661 by the Italian biologist Malpighi (1628–1694; cf. p. 205).

With the discovery of the circulation of the blood, Galenic physiology should, logically, have been replaced by new concepts. In fact it was not. Harvey, like Copernicus, disliked novelty for its own sake; having tidied up the relations between heart, veins, arteries, and blood he was quite satisfied to leave physiology where he had found it. He was content to have demonstrated that the heart is the most important organ of the body, "the sun of its microcosm"; he had no wish to alter the traditional roles of blood and spirit in animal nutrition. His was primarily the anatomist's point of view. Having solved the anatomical puzzle that had plagued his predecessors since Vesalius, he felt that he had done enough, and turned to embryological studies. The basic concepts of physiology awaited a century's further anatomical study before they experienced real modification.

Nor did the discovery of the circulation of the blood alter medical practice. Harvey remarked that the fact of circulation explained why an animal could so quickly bleed to death through a cut artery, but he did not therefore recommend discontinuance of phlebotomy, which lasted as standard therapeutic practice until the mid-nineteenth century. The existence of the circulation did explain why a poisoned or infected limb or organ could cause death, but the explanation suggested no cure. Changes in medical practice continued to come from quite different sources, from new diseases and new drugs, as they had been doing since the early sixteenth century, and the latest triumph of anatomical investigation produced no more effect upon medical practice than the first feeble attempts after 1500.

Though medical practice benefited so little from anatomy, the case was different with surgery. Physicians discovered new anatomical facts; surgeons made use of these facts, when they became aware of them. The separation by education and training of physicians and surgeons remained throughout the sixteenth and seventeenth centuries, though the education of surgeons improved continuously. Yet the physician despised the surgeon less than he had done, and was ready to impart his own new-found anatomical knowledge. In all countries lectures on anatomy for surgeons were organized; they were delivered by physicians, usually university M.D.'s, generally in the vernacular so that all might understand. And the surgeon could make use of anatomical knowledge as the physician could not. Especially was this true of the army surgeons, who were in ever-growing demand in the sixteenth century (except among the English, where their place was taken by naval surgeons); the battlefield proved an excellent training ground and all countries produced some notable surgeons in the sixteenth century.

Army surgeons dealt with two separate problems in the sixteenth century, and most successful surgeons wrote about them at length. The problems were the treatment of gunshot wounds, and the cure of new diseases rampant among the troops. In 1514 an Italian, Giovanni da Vigo (1460–1525), alarmed by the number of apparently poisoned wounds (probably infected with tetanus, much more common with bullet than with arrow or sword wounds), advocated cauterization of wounds made by lead bullets; lead was, of course, known to be poisonous if taken internally. Twenty years later, the French surgeon Ambroise Paré (1510–1590) lay awake all one night worrying because, in the aftermath of battle, he had run out of hot oil wherewith to perform what he took to be the necessary cautery. Finding that his patients treated with soothing lotions did better than those with cauterized wounds, Paré began publicly denouncing the practice of cautery, and advocated instead letting nature take its course. Cheered by his successes in (relative) innovation here, Paré searched for new methods in other areas of surgical technique: in the treatment of fractures and dislocations, of hemorrhages in amputation, of burns, and of obstetric practices, in this case continuing a tradition begun by Vigo.

Vigo's innovation in the treatment of gunshot wounds was destined to be discarded with scorn; his ideas on the cure of disease were sounder. The surgeon did not normally treat disease, but as there were no army physicians, army surgeons had to deal with the epidemic diseases which so frequently visited armies encamped in the field or laying siege to towns. Typhus, for example, became endemic whenever armies took to the fields, as it has remained ever since. But a more important disease, more devastating and more mysterious was that variously called "The new disease of the armed forces," "The Neapolitan disease," "The French disease," and "lues venerea"; it became known as "syphilis" after Fracastoro (1484–1553), an Italian humanist, physician, and astronomer, described its symptoms in classical Latin verse, complete with an allegorical tale of an Arcadian shepherd, Syphilis, who offended a god and was punished by a new disease. Contemporaries generally regarded it as American in origin, and there is good evidence that it was indeed present among Columbus' sailors on their return from the New World in 1493. There was a severe outbreak among the French soldiers laying siege to Naples in 1495, whence the disease spread in epidemic form all over Europe. Vigo was one of the first to use mercury to treat the disease. Mercurial ointment had been used since antiquity for skin diseases, so that it was natural to use it on the lesions of the skin characteristic of one state of the new disease; its use internally became specific for syphilis. Not all surgeons by any means approved; mercury produced distressing side effects, like excessive salivation; enormous quantities were necessary, and not all patients could endure the rigor of the treatment. Arguing that since the disease came from the New World, the cure must also be found there, many physicians and some surgeons concluded that the best remedy was guaiacum, a "holy wood" from South America; there were grateful patients ready to testify to its efficacy. (Either they were not suffering from syphilis, or the disease in these cases was already about to pass into a quiescent stage, for guaiacum lacks all medical value.) The case histories of surgeons like Paré, or the Englishman William Clowes, are filled with the results of the mercury-guaiacum battle; sooner or later most patients eventually died of the disease in any case.

The debate over the efficacy of mercury was complicated by the new school of chemical medicine, effectively founded by Paracelsus (1493–1541). Some chemical remedies had long been part of the pharmacopoeia: mercury, antimony, and sulfur ointments were recommended by Dioscorides and described by Pliny. Gold and iron were prescribed internally by Greek physicians: the reasoning behind their use was no doubt magical, but the iron at least may have done some good, while the gold can have done no harm to the stomach, whatever it may have done to the purse. A new drug was introduced into Europe with distillation in the late thirteenth century: alcohol, known for centuries as spirit of wine.

Fifteenth-century writers had enormous faith in the power of alcohol to preserve health, a faith commemorated in the names *eau-de-vie* and "aquavit." Cordials were alcoholic solutions in which herbs, fruits, or flowers had been steeped; as their name suggests, they were thought to strengthen the heart, and they were widely advocated in cases of plague. The next step was the preparation of essences: various forms of vegetable substance, steeped in water and distilled. Though the origin of these drugs was still herbal, they were prepared by chemical techniques and so in effect were themselves chemical. They prepared the way for the ingestion of mercury in cases of syphilis as well as for the introduction of antimony, mineral acids, metallic salts, and a whole new materia medica.

Chemical remedies did not acquire a respectable place in the pharmacopoeia without a struggle. Their advocates, like their opponents, were for long violently partisan, and the whole issue became clouded with the question of how vigorously a drug should work. For chemical remedies wrought far more powerfully than vegetable ones: a glass of wine which had stood in an antimony cup over night was a more violent emetic and purge than a large draft of most herbal concoctions, and it was easy to see the dangers possible in careless administration of such drugs. The greatest propagandist for chemical remedies, Paracelsus, combined his advocacy with violent denunciations of established medical practice (he once publicly burned several books by Galen, to the delight of the students at the University of Basle), an almost unreadable style (though he knew Latin, he chose to write in a mixture of Swiss-German and Latin that is painfully difficult to understand), and a profoundly mystical theory of disease. Yet Paracelsus was a most successful self-advertiser, and in the late sixteenth and early seventeenth centuries numerous medical writers styled themselves Paracelsians and advocated an ever-wider range of chemical drugs. The official pharmacopoeias yielded slowly; the influential Augsburg pharmacopoeia, in spite of an earlier ban on chemical remedies, introduced them in the edition of 1613; the French authorities forbade the use of antimony in the early seventeenth century, only to yield to public demand and rescind the interdict twenty years later; English physicians accepted most chemical remedies by the middle of the century. Times of plague produced occasional outbreaks of complaint by physicians who felt that official bodies were too conservative, but essentially the question was settled by 1650 and chemical drugs reposed on the apothecary's shelves side by side with the old "Galenicals." The gain to the patient is doubtful; the gain to chemistry was enormous, for apothecaries needed lectures and recipe books to teach them the art of preparing the new drugs, and demand soon produced the supply.

Chemistry also suggested new theories of disease as, curiously, the truly important discoveries in anatomy and physiology had not done. The only development along these lines in the sixteenth century was Fracastoro's interesting but unproductive theory of contagion, clearly an outcome of his investigations of syphilis. That some diseases were infectious —that is, that patients suffering from a particular disease often, if not invariably, communicated it to others—had been known since antiquity: Hippocrates, for example, had recognized tuberculosis as a contagious disease. Leprosy had been treated by isolation throughout the Middle Ages; the effectiveness of the policy is indicated by the fact that the disease had become rare in Europe by 1600. Quarantine—initially isolation for forty days—was introduced to combat bubonic plague in the fourteenth century. Now new contagious diseases, syphilis and typhus especially, appeared to complicate the picture. Fracastoro, in a kind of Lucretian theory, imagined that disease might be communicated by "seeds of contagion," disease particles of various kinds, some of which acted by being passed from person to person by direct contact, some of which left traces in clothing and bedclothes and so spread the disease, and some of which acted "at a distance" (these were involved in such diseases as ophthalmia, which is actually spread by insects). Fracastoro endeavored to explain why some people appeared immune to particular diseases, while some were extremely susceptible, by assuming that a force of attraction was at work between "seed" and individual; indeed the whole theory is appended to a treatise *On Sympathy and Antipathy* (1546), contagion being instanced as one example among many in nature. It was an admirable theory, but once again it could not influence medical treatment, nor was there any method of the time which could either prove or disprove its validity.

Chemistry, which was rapidly altering the drugs prescribed in medical treatment, was also the basis for a new theory of disease. This theory, adumbrated by Paracelsus, was dominant for most of the next century. Its basis was the belief that the body was like an alchemist's laboratory, the physiological processes involving alchemical transmutation. Paracelsus believed that chemistry offered a clue to the workings of nature because all natural processes producing changes in matter—like smelting, cooking, or brewing beer—were essentially chemical. He attached a mystic importance to fire; as he wrote "[God] enjoined fire, and Vulcan, who is the lord of fire, to do the rest. . . . this is the office of Vulcan, he is the apothecary and chemist of the medicine."[6] Vulcan is a kind of lesser demiurge, controlling the macrocosm; the microcosm, man, who partakes to the full of the chemical nature of the universe, has an inner Vulcan, which Paracelsus called the *archeus,* the presiding alchemist of the body. The archeus resided in the stomach, the central point for control of the whole physiology of the body. Paracelsus differed violently and explicitly from Galen and Aristotle over the nature of digestion: it did not, he insisted, involve mere *coction* (cooking) but an alchemical transmutation. (Needless to say, the insistence on the analogy between chemical reaction and human physiology does not make Paracelsus a forerunner of the nineteenth-century mechanistic school of physiology, for his chemistry was itself vitalistic.) The alchemist, Paracelsus pronounced, since he knows the body to be chemical in its nature, should devote his energies to preparing *arcana,* chemical remedies to combat disease by restoring the alchemical balance manifestly upset by disease.

It did not occur to Paracelsus that the alchemist ought to find out more about the physiological processes, as well as the chemistry of metals, before feeding the sick strong poisons like mercury and antimony. Nor was any such idea common among his immediate followers, the spagyrists or iatrochemists (medical chemists) of the late sixteenth and early seventeenth century. These spagyrists accepted wholeheartedly the role of the archeus and the concept of alchemy as the key to physiology; the mystics then proceeded to dithyrambic praise of the marvels to be expected if more physicians would become followers of Paracelsus, while the more rational concerned themselves with the preparation and use of chemical drugs. So for example "Basil Valentine" (the pseudonym of Johann Thölde) called the first detailed study of the chemistry of antimony and its compounds *The Triumphal Chariot of Antimony* (1604); though there is a good deal of highly allegorical language in this influential work, it is quite possible to understand and follow most of the preparations, and the entities named are genuine chemical substances. When, in the first quarter of the seventeenth century, chemical drugs became a necessary part of official pharmacopoeias, a series of chemical teachers began an influential series of textbooks. Some lectured to apothecaries; their books dealt with the best ways to prepare standard drugs. Others lectured to medical students; their books naturally stressed the use of the drugs which the physician would obtain from the pharmacist. On the whole, the French leaned towards the pharmaceutical side; witness for example the enormously popular *Elements of Chemistry* (1610) of Jean Beguin, which was used throughout the century. The Germans tended to deal with medical prescriptions; an example is Oswald Croll's *Basilica Chymica,* (*"Royal Chemistry,"* 1609) which assumes that others have already prepared the drugs whose use he describes; this was also a widely read work.

Only one chemist of the seventeenth century followed the more difficult road suggested by the doctrine of Paracelsus and tried to understand the workings of alchemical physiology. This was J. B. van Helmont (c. 1577–1644) whose collected works, *Ortus Medicinae* (1648, translated into English in 1662 under the title *Oriatrike or Physick Refined*) is a strange mixture of mysticism, medicine, and chemistry. Helmont grasped the fact that chemistry was capable of becoming a science; he was, however, remote from the spirit of his age, too much "a philosopher by fire" to conceive a rational framework, and his is a mystic science. So far did he go in his chemical physiology that he derived his chemical concepts from the workings of the human body rather than the reverse. Having become aware that some bodily secretions connected with digestion were acid (at least to taste) and some apparently alkaline, he concluded that digestion must be a fermentation occasioned by an acid-alkaline reaction in the stomach. He had studied the neutralization of acids in vitro; it happened that the most common alkali (potassium carbonate, then known as potash or salt of tartar) effervesces with acids, emitting carbon dioxide; Helmont concluded that the same was true of the reaction in the stomach. (He obviously had good supporting evidence for this belief.) This acid-alkali fermentation was only one example of what he thought were the changes produced by ferments (which he likened to yeast) in the various parts of the body; all were in charge of an archeus. Yet, mystic though he was, Helmont had a firm grasp of many chemical concepts, and his theory was suggestive.

Though chemical advances—notably the work of Robert Boyle—showed difficulties in the acid-alkali dichotomy, the acid-alkali theory of digestion persisted and was carried into other areas of medicine by ardent followers of Helmont. To these Helmontians an acid-alkali imbalance, rather than an unbalance of humors or spirits, was the cause of disease. This suggested nothing new about the disease itself, but much about treatment. The administration of mild alkalies (like sodium bicarbonate) long antedates the nineteenth-century discovery that the principal ingredient in gastric juice is hydrochloric acid. (This, under the name "spirit of salt" was one of the new mineral acids in Helmont's time; one of his younger contemporaries Johann Rudolph Glauber [1604–1670] soon after Helmont's death suggested its use in softening tough meat before cooking.) The persistent attempt, prolonged for over a century, to dissolve urinary calculi *in situ* was related to the acid-alkali theory. So too was the preparation and administration of a host of prescriptions involving either invigorating acids—vinegar, lemon juice, dilute mineral acids—as for example in scurvy; or soothing alkalies, often in the form of the powdered shells of crustaceans. Meanwhile the pharmacist's shop became more and more a chemist's shop, as the new chemical drugs gradually became more popular than the old herbal remedies, and the pharmacist himself became ever more skillful a chemist.

IV

The Scientific Revolution

Chapter 9

THE UNIVERSE OF MATTER AND NUMBER

In the fifteenth century Italy was the most prosperous and cultured region of Europe. There the Renaissance of art and literature, of science and technology, flourished first and most brilliantly. Enjoying such an inheritance, the Italian universities of the sixteenth century excelled all others in the quality of their teaching and the originality of their faculty. And among the Italian universities Padua was outstanding: it was Padua that attracted Vesalius, Fracastoro, Fabricius, Harvey, and above all Galileo, who taught there from 1592 until 1610. The medical school at Padua was so famous that it drew students from all parts of Europe, yet the university was no less remarkable for its philosophers who pursued the medieval investigation of the logic of experimental science. So one of them, Jacopo Zabarella (1533–1589), clearly explained how hypotheses are formed by investigating phenomena, and verified or refuted by discovering whether the effects deduced from them are found or not:

> . . . when we form some hypothesis about the matter we are able to search out and discover something else in it; where we form no hypothesis at all, we shall never discover anything. . . . The other help, without which this first would not suffice, is the comparison of the cause discovered with the effect through which it was discovered, not indeed with the full knowledge that this is the cause and that the effect, but just comparing this thing

with that. Thus it comes about that we are gradually led to the knowledge of the conditions of that thing. . . .[1]

While the Paduan doctrine anticipated much that Francis Bacon and Galileo were to teach later, the burning issues of the early sixteenth century were to be found, not in the experimental sciences, but in astronomy. The availability of Ptolemy's *Almagest* in Greek had not removed all problems, as the pioneers of Renaissance astronomy, Peurbach (1423–1461) and Regiomontanus (1436–1476), had hoped it would. Ptolemy's system was mathematically complicated; it did not agree with the physical picture of the universe provided by Aristotle; its tables contained errors (for predictions went astray by days or weeks); and above all it failed to teach how the notorious error in the calendar might be rectified. This last problem, of so great concern to the Church, caused the papacy itself to bless the search for a new system.

Several new systems were proposed in the early sixteenth century: Fracastoro, for instance, revived the homocentric spheres of Eudoxos. The one that endured was the work of an astronomer deeply learned in Ptolemy and mathematics, and profoundly influenced by the Italian universities. Nicholas Copernicus (1473–1543) enrolled at both Padua and Ferrara, but it was Bologna—celebrated for its astronomers—that influenced him most, through his work with Domenico Maria Novarra in the years 1496–1500. It can hardly be too much emphasized that Copernicus was not a logician, nor a philosopher, but a mathematician. He was scarcely even an assiduous observer of the heavens, for he left the determination of the celestial motions much as he found it. Only brief parts of *On the Revolutions of the Celestial Orbs* (1543) were devoted to the justification of Copernicus' triple motion of the Earth in philosophical terms; Copernicus' strength lay in his mastery of celestial geometry and on this achievement he rested his case. The bulk of *De Revolutionibus* is Ptolemaic astronomy stood on its head, the work of a master technician steeped in every detail of the *Almagest*. Copernicus had perceived that one orb given by Ptolemy to each planet and to the Sun did no more than reflect the true motion of the Earth; how much more simple and harmonious to abolish these superfluous orbs by letting the Earth revolve around the Sun, in its proper place between Venus and Mars! From this change, essentially for him a *geometrical* change, Copernicus thought there followed a theory that was credible because it was apt, consistent, and orderly, whereas in his eyes all geocentric systems were misshapen, inconsistent, and inharmonious, and consequently false.

Only the first step to a scientific revolution was taken by Copernicus, despite the magnitude of his daring assumption that the Sun stands still. Like him, few who participated in the greatest revolution in thought that has ever occurred saw more than a step or two ahead. Like him, all of them were affected by a sense of the sterility of late-medieval scientific thought, strengthened by the neoclassicism of the Renaissance, and there were others besides Copernicus who found an escape from this sterility by returning to half-forgotten ideas of antiquity. Many, too, like Copernicus were guided by a renewed vision of a world-order that should be Platonic rather than Aristotelian in character, one characterized by harmony and founded upon orderly mathematical relationships. Medieval astronomers, Islamic and Christian, had been content with any geometrical model that would "save the phenomena," that is, would grind out numbers roughly in accord with observation. Copernicus asserted that all such models were figments; the true geometry of the heavens would be known by the "unalterable symmetry of its parts," and by the contrast between the evident necessity of its relationships and the arbitrary character of false systems.

Here there are already signs of new trends that were to lead far. Copernicus' world was still the closed world of the Middle Ages, encircled by the distant sphere of the fixed stars, now indeed more remote than before; for him all the crystalline orbs were still real. Yet his astronomical system was to demand a new cosmology and a new physics, for the motion of the Earth shattered Aristotle's categories and his whole system. No more could "up" and "down," "light" and "heavy," be referred to a fixed central Earth; the dichotomy between the perfect heavenly bodies and the corrupt sub-lunary region was destroyed when the Earth itself became a planet, and the nature of motion itself was changed. It was far less significant that man's home ceased to be the hub of the universe (for that had always been seen as the sink of imperfection and an insignificant speck in the cosmos) than that the whole intellectual order was thrown off its bearings.

How much of this Copernicus perceived is not clear. He made little attempt to adjust cosmology and physics to his new system. Awareness of the certainty of ridicule, he admitted, had kept his ideas private for more than thirty years until the close of his own life, but if he felt the potential weight of more solid criticism he gave little hint of it by offering any defense. Such defense as he gave consisted of turning his potential opponents' arguments against themselves: the motions of the Earth are natural, therefore they cannot cause its disruption, any more than the motions of the heavens disrupt them; rising and falling, being merely relative to the Earth, occurs as before; circular motion belongs to spherical bodies, therefore it is proper to the Earth. Only when he wrote of gravity did Copernicus do more than this, and then he omitted to call attention to the way in which his differed from the traditional notion in which gravity was a function of position; gravity is, he wrote, "but a natural inclination, bestowed on the parts of bodies by the Creator, in order to combine the parts in the form of a sphere and thus contribute to their unity and integrity. And we may believe this property present even in the Sun, Moon, and Planets. . . . ."[2] This utterly un-Aristotelian and unscholastic statement, attributing to matter universally a mysterious inclination or force, pregnant for the future as it was, Copernicus neither justified nor explained. Yet without this principle, which places him with the moderns, Copernicus' system would have been an absurdity.

The technical novelties of this system were few, given the change of focus from Earth to Sun.* Copernicus' universe is pure Greek, the world Ptolemy would have made had he followed Aristarchos. Modeling himself on Ptolemy, Copernicus recomputed all the elements of his universe from observations (mostly Greek, a good number Arabic, a few of his own), reaching results that correspond closely to those of the *Almagest,* save that now the relative sizes of the orbs were given for the first time. As a natural consequence of the adaptation, the order of the planets (from the Sun) became Mercury, Venus, Earth, Mars, Jupiter, Saturn—the order of increasing periodic times (Figure 6). Believing that Ptolemy's equant was a false device, because it made circular motion nonuniform about the geometric center, Copernicus was compelled to introduce a number of smaller spheres to account for the apparent variations in the planet's speed as seen from the center of its sphere. Thus the Copernican system—which found the motions of the Moon and of Mercury intractable, as all systems have done—was a good deal more complicated than a nest of six orbs interposed between

* More exactly, the center of the Earth's orbit, for Copernicus' was not a *literally* heliocentric system but a heliostatic system.

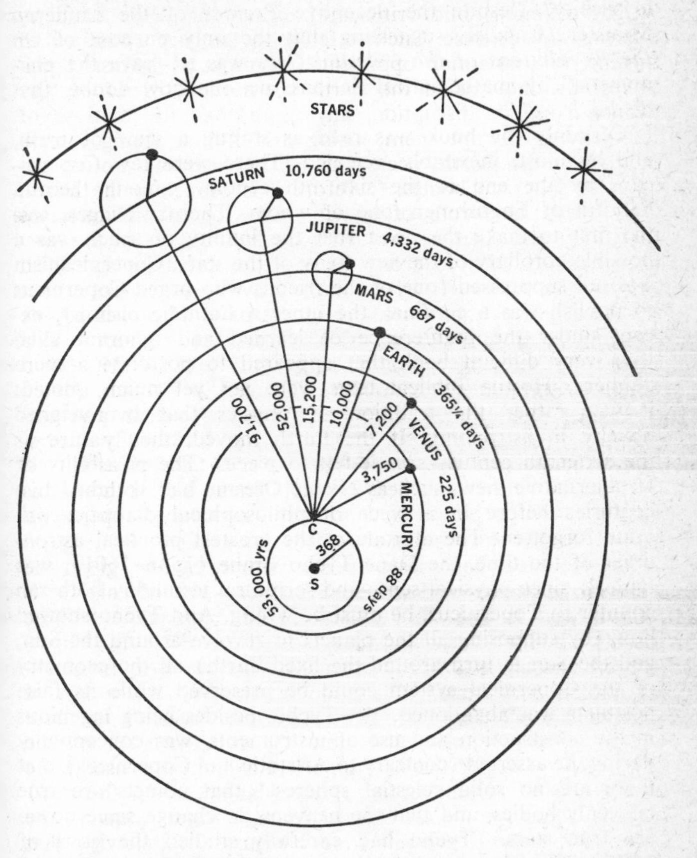

Fig. 6. The Copernican System, much simplified and not to scale. *S* represents the Sun, *C,* the center of the Earth's orbit. The orbits for Mercury and Venus are only approximately indicated.

the Sun and the fixed stars. Nevertheless, Copernicus could claim that it was a great deal more symmetrical than the Ptolemaic: in particular it did away with the circles reflecting the Earth's motion that Ptolemy had necessarily introduced. That it corresponded equally well with observation Copernicus could easily show. Did he believe that it also corresponded to reality? Despite the deceptive *Preface* of the Lutheran Minister Osiander, suggesting that the only purpose of *On the Revolutions of the Celestial Orbs* was to "save the phenomena" by mathematical fictions, no one now doubts that he did.

Certainly the book was read as stating a view of truth, and by most, inevitably, rejected. There were few Copernicans by the end of the sixteenth century, among them a handful of Englishmen, one of whom, Thomas Digges, was the first to make the point that the infinity of space was a possible corollary of the new fixity of the stars. Copernicanism was not suppressed (one of the friends who urged Copernicus to publish was a cardinal, the other a Catholic bishop), except under the indifference of learned and ignorant alike to a very difficult book that appeared to postulate a mere chimera. Hostile Biblical texts were not yet much quoted; it was rather the tradition of physics that overweighed novelty in astronomy. If the Earth moved, then nature as the sixteenth century saw it fell to pieces. The possibility of an alternative view, such as Nicole Oresme had sketched two centuries before in a piece of philosophical dialectic, was quite forgotten. The attitude of the greatest practical astronomer of the time, the Dane Tycho Brahe (1546–1601), was typical: since physical sense and scriptural testimony both ran counter to Copernicus, he must be wrong. And Tycho showed how (by supposing all the planets to revolve around the Sun, and the Sun in turn around the fixed Earth) all the geometry of the Copernical system could be preserved while its false postulate was abandoned. Yet Tycho, besides being ingenious in the construction and use of instruments, was conceptually daring; he asserted (contrary to Aristotle and Copernicus) that there are no solid celestial spheres,* that comets are true heavenly bodies, and that the heavens do change since novae are true stars. (Tycho had carefully studied the nova of 1572.) For the moment the geostatic "Tychonic system" seemed blessed with all probability. Its downfall was still, when its author died, concealed in Tycho's own notebooks and the mind of Johann Kepler.

For a change began towards the end of the century. Compromising with tradition, the older Copernicans, like their master, had tried to adapt the Earth's motion to physical theory; their successors had begun to say that if Aristotle's

* The abandonment of solid spheres, moreover, raises difficulties as great as those which Tycho objected against the motion of the Earth.

physics stood in the way of the heliostatic system, then it must be replaced by a more amenable physics. Giordano Bruno (1548–1600) was already broadcasting a new scientific metaphysics, apt for an infinite universe and a plurality of worlds. Galileo Galilei (1564–1642) was working his way from the idea of impetus to a new dynamics, which would explain how the Earth might move. William Gilbert (1540–1603), the English philosopher of magnetism, having demonstrated that the Earth is itself a great magnet, claimed that its axial rotation was a necessary physical consequence of its magnetic sphericity. (Gilbert, however, rejected the second motion of the Earth, its annual revolution about the Sun.) Most important, Kepler (1571–1630), not content to ask whether the Copernican system might be true, sought to discover why it must be true.

Kepler's search, the dominant theme of his life, embraced a strange combination of Platonic and physical speculation. He desired to find in the architecture of the heavens both an obvious mathematical harmony, and a physical explanation (actually conceived in Gilbertian terms) of why this harmony exists. Thus Kepler was both the last in the line of pure geometrical astronomers who sought to define the celestial motions by lines and curves, and the first to conceive of a celestial mechanics that was not a mere hypostatization of geometry. In his first book, however, *The Cosmological Mystery* (1596), the former theme is dominant; here he tried to prove (but knew he had not) that the sizes of the six planetary orbits corresponded to those of the five regular geometrical solids, fitted one within the other in a certain order. An absurd enough notion, yet the idea that there is some correlation between the sizes of these orbits was a sound one. Twenty years later (in the *Harmony of the World*, 1619) Kepler found it; the cubes of the diameters of the orbits are proportional to the squares of their periodic times.

From this, Kepler's Third Law of planetary motion, it follows that many solar systems would be possible, not merely one as Kepler thought in 1596. Since then his thought had matured, and he had gained possession of the observations of Tycho Brahe. During their brief association in Prague before the latter's death, Tycho had set Kepler to work upon the classical problem in astronomy—the determination of orbits. Fortunately beginning with Mars (whose orbit is the most eccentric of all, except that of Mercury) Kepler found first that the planes of all the orbits passed through the Sun, confirming his attribution to it of a physical significance as the center of the system, which for Copernicus the Sun had not possessed. Then, considering the velocities of a planet at different points of its orbit, he discovered his Second Law: that the radius vector between the Sun and the planet sweeps over equal areas of the orbit in equal times. This relation was of great use to him in subsequent analyses. So far Kepler, like all his predecessors, had thought of the orbit as circular; anything else was inconceivable. Seeking now to define this circle, and particularly to find a center of uniform motion within it, he found that the Second Law, the circular orbit, and the observations were mutually incompatible. After years of vain computations he stumbled upon the answer, the so-called First Law: the orbit was an ellipse with the Sun at one focus. All this work Kepler set forth in his *New Astronomy, or Celestial Physics* (1609).

Throughout these long analyses upon a multitude of useless hypotheses Kepler was constantly influenced by the necessity to supply each with a plausible physical basis. In good Aristotelian fashion he imagined the planets to be driven by a force radiating from the rotating Sun; as this decreased with distance, the outer planets were driven more slowly. To account for their elliptical orbits Kepler postulated a polarized magnetic force between the Sun and each planet; in the aphelion half of the curve the planet was repelled from the Sun; in the perihelion half (turning the other pole toward it) it was drawn back again.

Few of Kepler's contemporaries understood his obscure, richly autobiographical writings: one of them was the English astronomer, Thomas Hariot (1560–1621), who was perhaps the source of the admiration felt for them by a group of his countrymen, including ultimately the young Isaac Newton. Kepler's difficult mathematics—as he remarked himself—deterred many; many also were repelled by his trust in numerical harmonies, and by his talk of attraction and repulsion which seemed to fit rather an animistic than a mechanistic view of the universe. For in Kepler's later years the mechanistic conception of nature with its comparison of the universe to a vast, complex machine, governed by immutable laws affecting the movement of every part, was rapidly commending itself to science. Kepler's laws were to be the foundation of celestial mechanics, and he himself tried to make them so, but this was clear to none. His mechanical hypotheses were immature. Gilbert had made terrestrial

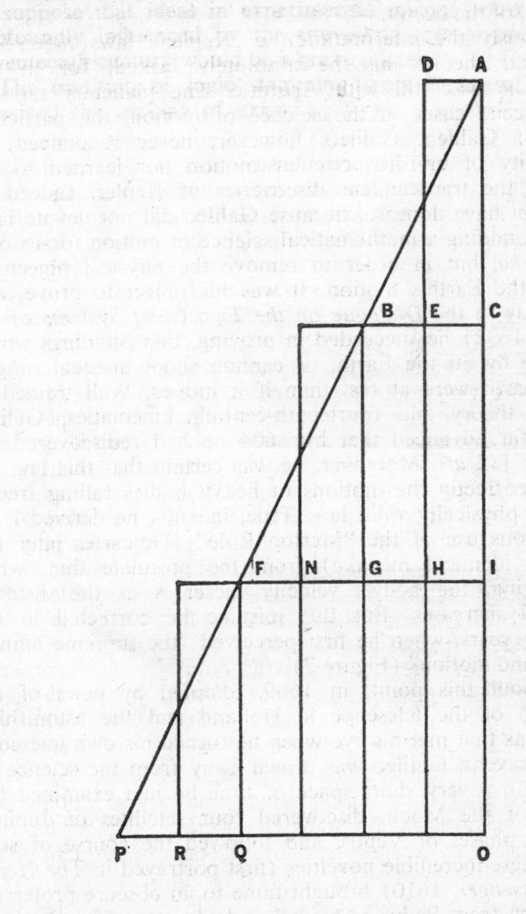

Fig. 7. Galileo's demonstration of the Law of Fall.
The lines *BC, FI, PO* represent velocities after the times *AC, AI, AO* to which the former are proportional. By the Merton Rule—previously proved by Galileo—the distances traversed in the equal time intervals *AC, CI, IO* are as the enclosed areas *ABC, BCIF, FIOP;* hence, as 1, 3, 5, 7, . . . or $s \propto t^2$.

magnetism respectable; in Kepler an astral magnetism smacked of the occult.

Obviously the interpretation of Kepler's laws required a dynamical theory that he absolutely lacked, for in this respect he was still with Aristotle. The planetary motions were special cases in the science of motion, the particular study of Galileo. Galileo, however, never renounced the superiority of uniform circular motion nor learned to appreciate the transcendent discoveries of Kepler. Indeed he could not have done so, because Galileo did not devote himself to founding a mathematical science of motion for its own sake alone, but in order to remove the physical objections against the Earth's motion. It was his object to prove, and eloquently in the *Dialogue on the Two Chief Systems of the World* (1632) he succeeded in proving, that buildings would no more fly off the Earth, or cannon shoot unequal ranges, if the Earth were at rest than if it moved. Well trained in impetus theory and fourteenth-century kinematics, Galileo had so far advanced that by 1604 he had rediscovered the law $s = 1/2\ at^2$. Moreover, he was certain that this law described correctly the motions of heavy bodies falling freely; it was a physically valid law. True, in 1604 he derived it by a fallacious use of the "Merton Rule" (Descartes later fell into the identical mistake) from the postulate that, when accelerating, the body's velocity increases as the distance traversed increases. But this mistake he corrected in the next few years, when he first perceived "the supreme affinity of time and motion" (Figure 7).

At about this point, in 1609, tempted by news of the invention of the telescope in Holland and the astonishing revelations that met his eye when he turned his own telescope to the heavens, Galileo was drawn away from the science of motion. In a very short space of time he had examined the surface of the Moon, discovered four satellites of Jupiter, seen the phases of Venus, and followed the course of sunspots. These incredible novelties (first portrayed in *The Heavenly Messenger,* 1610) brought fame to an obscure professor, and return from Padua to his beloved Florence. They brought time-wasting controversy too. His priority was impugned, his observations were assailed, and the use he made of them to exalt the Copernican over the Ptolemaic system was severely criticized.* Galileo discovered that he had facility and cutting

* Not to labor the obvious, it may be noted that all Galileo's points against the geostatic hypothesis are good against the Ptolemaic but not against the Tychonic system, for the latter is completely equivalent to

power as a debater, which he expended freely over the next twenty years in defense of Copernicus. Because others besides Galileo had taken the offensive, and some (not Galileo in the first instance) had raised the question whether the Bible was to be taken literally as a textbook in physics, the Catholic Church and its preachers had become aware of a danger to theology. Inevitably Galileo was drawn into this argument too; inevitably he was led to distinguish between authority in religion (faith) and in science (reason). He appealed to a venerable tradition in asserting that the Bible sometimes spoke metaphorically. Correct, replied the theologians, but we only admit this when the literal interpretation is demonstrably false, and you have not proved that the Sun does not move.

To Galileo, a declaration of his Church against Copernicus (against scientific truth as he saw it) would be a disaster. It came, though mildly, in 1615. He tried in vain to reverse it; his last resource was to take up the theologians' challenge and prove the motion of the Earth from his theory of tides, after demolishing every objection. He wrote the *Dialogue* of 1632. It came too late. The Church was committed to a policy which compelled it to bring Galileo to trial, submission, and sentence in 1633.

Galileo's case in the *Dialogue* rested partly on the qualitative observation of the sky with the telescope which he had made since 1609. Extracting its logical force with consummate skill, he destroyed the Aristotelian cosmos; the Moon, Sun, and Planets were reduced to the status of physical bodies like the Earth. The sublunary world and the cosmos were made one—or nearly so. For the first time it was clearly announced that the whole universe and each part of it are subject to the same laws, though Galileo wrote of reasoning rather than law. Galileo's is a universe whose reality consists of matter and motion only; the qualities of bodies (taste, color, texture, and so forth) are mere secondary phenomena arising through sense perception of these realities. It is, moreover, a universe of number:

> The book [of nature] cannot be understood unless one first learns to comprehend the language and read the letters

---

the Copernican. This Galileo knew, and therefore he always attacked the Ptolemaic astronomers and the Aristotelian philosophers. He confessed that he had only one *proof* of the motion of the Earth, which he evolved from a theory of the tides which was (in fact) erroneous.

> in which it is composed. It is written in the language of mathematics, and its characters are triangles, circles, and other geometric figures without which it is humanly impossible to understand a single word of it. . . .[3]

Despite the misleading testimony of the senses (Galileo was keenly aware of the perils of naive empiricism) the physicist's universe is the true universe; and for his description of this universe the physicist must employ quantitative, mathematical language. Galileo insisted that mathematical physics does not apply to some ideal (Platonic) or arbitrary (medieval) state of affairs, since, provided the mathematical picture is complex enough, the physicist can make it correspond to reality as exactly as he pleases. In the last analysis, therefore, physical curves (trajectories, hanging chains, optical figures) are geometric curves; actual bodies—when we know how to allow for their material content—behave like geometric solids; empty space has the properties of Euclid's geometry.* There are no distinctions in the universe, no privileges, no peculiar regions to obstruct the rule of geometry.

In such a universe the geometry of motion, kinematics, is supreme. In the *Dialogue* Galileo, sketching his new kinematics for the first time, made it the main base for his attack upon the traditional cosmology. With many analogies from experience he explained how a body moving upon an infinite, frictionless plane would continue to move indefinitely at the same velocity. From this followed easily, with a further wealth of illustration, the major principle of his argument: if an observer relates events to certain coordinates, with respect to which he considers himself at rest, he is unable to determine from these events whether he himself and the coordinates are at rest, or participating in a common uniform motion. The observer upon the Earth, accordingly, taking the Earth as providing his coordinates, can never tell whether the Earth moves or not; the terrestrial events (and by an obvious corollary, the celestial events) are the same in either case. From this reasoning it was clear (as Newton saw) that any statement about absolute motion must postulate the existence of fixed and absolute space-time coordinates.

Perhaps this might be called Galileo's "cosmological principle," since it complements in kinematics the assertion of

* Galileo would have entertained no notion that space might be filled with forces or fields.

uniformity which Galileo had already made for geometry. He stated it very imperfectly, or rather, overstated it; he should have declared for a "common, uniform, *rectilinear* motion." But—victim of a curious mathematical fallacy—Galileo could never distinguish between a circumference of large radius and a straight line. Betrayed by his particular view of the infinite and the infinitesimal, he thought that a planet could revolve forever *in a perfect circle* impelled by its inertia alone. Indeed, the concept of inertia, as Galileo understood it, provided the physical explanation of the planetary motions, now performed without the aid of solid orbs. Galileo could never accept Kepler's discoveries (since they conflicted no less with his physics than with Aristotle's), nor could he have come near to understanding central forces and gravitational attraction.

In their separate spheres Kepler's new astronomy and Galileo's kinematics, which he both justified and largely amplified in relation to terrestrial physics in the *Discourses on Two New Sciences* (1638),* could work well independently; in conjunction they produced a hopeless scientific enigma. The predictions yielded by each theory were separately verifiable, but together they made no sense. In the event, while Galileo's ideas were quickly appreciated and his law of fall ($s = 1/2\ at^2$) used everywhere, Kepler's ideas were neglected. Even his descriptive laws were ignored. A reconciliation awaited Newton.

Meanwhile one of the greatest of pure mathematicians set aside both these mathematical physicists in pursuit of the mechanistic ideal in its purest form. For René Descartes (1596–1650) pursued yet a third independent path. Galileo's kinematics he scorned because Galileo had not (in his view) understood the nature of either gravity or motion. Kepler's laws he knew but never applied; Kepler's celestial physics he dismissed as wholly unphilosophical. And though Descartes performed one considerable feat of mathematical physics, his *Dioptrique* (1637) in which he first announced the sine law of refraction, he constructed a complete physics and cosmology that lacked all mathematics. Like Aristotle, Descartes sought first for indubitable axioms. The first of these were necessarily metaphysical, the certainty of his own existence and of God's. The next was a principle of truth: all

* Kinematics and the strength of materials, but there is much else studied in this great book belonging to other branches of physics (light, sound, the vacuum, motion of pendulums, etc.).

ideas that are clearly and distinctly perceived are true. Then Descartes found the two physical axioms that satisfy this criterion: his definition of matter as extension in space, and the impossibility of the vacuum (for a space empty of matter would be a contradiction). Finally he defined the laws of motion.

So far Descartes in the *Principles of Philosophy* (1644) had proceeded, as he thought, like a geometer proposing the axioms of a formal science. But the world that he derived from them was as imaginary as that of Aristotle, and in some respects more arbitrary. The details may be passed over. Descartes' is the first cosmos within which the solar system is but one member; the Sun is situated at the center of a rotating vortex of aethereal matter bearing round the planets, and all the stars are centers of precisely similar vortices. It is also an evolutionary cosmos, in that planets are formed from decayed stars, while terrestrial matter, geological formations, and so forth arise from the further decay of planets. It is a cosmos composed of nothing but matter in motion, all whose events are caused by the impact of moving particles upon one another; heat, light, magnetic force, the growth of plants and every function in physiology (except those controlled by human will) are interpreted as special cases of this dynamic action. The spaces that appear vacuous are readily crossed by interparticulate actions because they are absolutely full of aether, an aether which is indeed the ultimate source of motion, and hence of all phenomena, since to it gross matter transfers its own movement and receives it back again. And in Descartes' cosmos there is no entropic degradation, for the "quantity of motion" is perpetually constant.

Despite the *ad hoc* character of Cartesian hypotheses—the type of hypothesis that Newton later declined to feign—his physical theory won immediate admiration. For two generations, at least, no scientist was outside its influence. Within a few years of Descartes' death a school of orthodox Cartesian scientists took shape in France and Holland which continued straightforward exposition and defense of his ideas until about 1750. Outside this dogmatic school, in England as on the Continent, almost everyone adopted more or less large portions of the *Principles,* though detection of its errors and inconsistencies also took place rapidly. The reason for this success is clear: Descartes had provided a view of nature, a mechanistic view without any concessions, such as the seventeenth century longed for. He had provided a theoretical model that fitted precisely the metaphysical position in which nearly all scientists, by the time of his death, felt themselves comfortable; one that abjured natural spirits and all occult relationships; that denied the possibility of having a priori knowledge of anything save matter, motion, and mathematical or logical relationships; and that asserted the complete, and resolvable, consistency of all causal relationships. Concealed though this was by the obvious antithesis of their ideas on particular points, Descartes and Galileo were completely at one, except that the latter had emphasized the mathematical architecture of nature, the former its mechanical architecture which he was (inevitably) forced to imagine. Only a point of view that was rapidly becoming outmoded and irrelevant, which still attached importance to occult, immaterial causes in physical phenomena (represented by a few German scientists and by the alchemists) stood completely aside from this increasingly coherent metaphysics. As it was realized in Descartes' model, this for a time promised all the intellectual satisfaction that Aristotle's physics had given to the Middle Ages. Many, indeed, felt (like Spinoza and Leibniz) that the structural elements of this model could require no empirical verification, since their truth was metaphysically established. Although the detailed explanations of phenomena in the *Principles* might need modification as experimental knowledge and mathematical analysis progressed, such displacements could be effected without altering the essential character of the Cartesian system.

This character was by no means incompatible with the development of mathematical and experimental researches in science; indeed, Descartes had admitted that many more such researches were required for the perfection of his system. So, for instance, the great Dutch physicist Christiaan Huygens (1629–1695), for many years the kingpin of the French Academy of Science in Paris and famous for his work in astronomy, optics, and mechanics, remained Cartesian always in his attitude to scientific explanation, though he could write of Descartes' *Principles* that it was "science fiction" (*un beau roman de physique*). In his view:

> . . . in the true Philosophy . . . one conceives the causes of all natural effects in terms of mechanical motions. This . . . we must necessarily do, or else renounce all hopes of ever comprehending anything in Physics.[4]

Against Newton, Huygens believed that gravity must be caused by the pressure of aethereal matter on bodies, for only in such a way could the laws of gravity be known by tracing them to mechanical principles, that is, to the impact of particle upon particle. Huygens did not live to see the destruction of his hope that Descartes' mechanistic physics and the rigorous theory of the science of mechanics would be reconciled. When the contradiction between speculative mechanism and mathematical mechanism became fully apparent, Cartesian physics was doomed.

This contradiction was fully worked out by Newton, in the context of the theory of gravitation. But Newton was, with reservations, a "mechanical philosopher," and to understand fully how Newton's theory of universal gravitation condemned Descartes' cosmos to oblivion, one must first examine the development of the mechanistic philosophy of nature.

Starting from the same metaphysical position as Descartes it was not difficult to reach ideas analogous to his but different from them: physical speculations in which the vacuum was allowed and the infinite divisibility of matter embraced by Descartes rejected. Galileo, for instance, was both an atomist and a vacuist. Later the atomic theory was taught by the philosopher Pierre Gassendi (1592–1655), adapting Lucretius' poem *De rerum natura* to modern knowledge. Most scientists found the dynamic theory of Descartes more convincing. On the other hand, the mechanism of Robert Boyle (1626–1691) was totally eclectic; he regarded the metaphysical arguments for and against atoms and void as far less significant than the general explanatory power of the "corpuscular" philosophy, analyzing phenomena in terms of parts, motions, and structure. Indeed, it was Boyle who first really brought out the usefulness of the concept that matter has structure, which he associated (for example) with the selective color reflection of objects. Boyle was also distinguished from the mechanistic philosophers of the Continent, though supported by many other Englishmen, in regarding as essential the empirical demonstration of the mechanistic hypothesis. Possibly microscopes would one day reveal the microstructure of bodies; until then, the inference from macroevents to microhypothesis must be as complete and rigorous as possible.

Apart from his experimentation (pp. 175–177; 191) Boyle made an impressive contribution to the mechanization of the universe by extending mechanistic theories to chemistry. He set himself to try

> whether I could by the help of the corpuscular philosophy . . . associated with chymical experiments, explicate some particular subjects more intelligibly, than they are wont to be accounted for, either by the schools or [by] the chemists.[5]

The chemists, in so far as they were not satisfied to pursue technique alone in the service of medicine or manufacture, had accounted for the changes accompanying a chemical reaction either by appealing to the Aristotelian theory of qualities, or by supposing that alterations in the proportions of the four elements had taken place. Boyle considered chemical change as purely mechanical. The existing arrangements of the component corpuscles (which might be either "primary" [atoms] or "secondary" [molecules]) were broken up in the reaction, and certain motions occurred by which the corpuscles were redistributed into new patterns. The former gave the reagents their characteristic textures and properties; the new patterns furnished the different textures, and therefore different properties, of the products of the reaction. Boyle was particularly interested in the cases where the reaction is reversible, as demonstrated for example by color changes produced when a solution is rendered either acid or alkaline. Here the change of color was clear evidence of a transformation of physical structure, and Boyle noted (in contradistinction to some earlier chemists) that *three* states were possible: acid, neutral and alkaline. Holding this mechanical view, which he applied with great success to a very large number of chemical experiments of which many were both original and ingenious, Boyle was highly doubtful of the concept "element" in its customary (ultimately Aristotelian) sense. For he believed that the number of types of ultimate particle (if there was more than one type) was small, and also that in theory—and perhaps at length in practice too—the secondary combinations of these ultimate particles into corpuscles could always be broken down. If one element—say water—could be broken down and recombined into earth (and Boyle thought he had experimental evidence for this), then there was no reason to apply the term "element" to anything but the ultimate particles themselves.

Boyle's physical chemistry—as it may well be called, for it was entirely a theory of structure—had a profound influence upon Newton, who was also experienced in chemical investigation. Newton added a new dimension to Boyle's theory by explaining the dissolution, motion, and recombination of corpuscles occurring in a chemical reaction in terms of relative forces. Obviously there is some force of cohesion between the particles of bodies, even in solution, a force that Newton supposed to be very powerful at minute distances. Now if the force varies from one type of corpuscle, or particle, to another so that the force between *A* and *B* is less than that between *A* and *C*, then if *C* can be dissolved in (say) the compound fluid *AB*, *A* and *C* will combine and *B* be precipitated. In this way Newton explained how metals precipitate each other from solution in a definite order. Both Boyle and Newton agreed that some combinations or textures of the particles might be so tight and highly interlocked that the substance so formed, whether natural (gold) or factitious (glass), could not be broken down by any known means. Neither qualified such a substance—which would always be recoverable from a reaction—as an element.

In chemical theory, as in all other respects, Newton brought to its fulfillment the work of those who had preceded him in constructing the mechanical view of the universe; yet at the same time he showed the insufficiency of that view, as they had held it. His theory of chemical force for example—which clearly owed much to his concept of gravitational force—undercut the more naive mechanistic hypothesis that reactions occurred by impact alone, and it stated the unsolved problem of the nature of this chemical force. Newton believed that a number of such forces must be inferred from the phenomena of nature, such as the force by which bodies bend light (in refraction, diffraction, and so on), and above all the force of gravity, whose mathematical theory he constructed completely. But he could not explain the nature of the gravitational force, only define it and infer the necessity for its existence.

No scientist has excelled Isaac Newton (1642–1727) in the combination of his abilities in experimentation, mathematical physics, and pure mathematics.* Although he was not remarkably precocious in youth, his creative ideas first took shape in his mind when he was still a very young man—as with all mathematicians. In 1665–1666, when he retreated to his home from a plague-infected university, he developed the

* Lucasian Professor of Mathematics in the University of Cambridge, 1669–1696, then Warden and Master of the Royal Mint, London; President of the Royal Society 1703, knighted 1705. He was also a classical scholar, a theologian, a chemist, an amateur of alchemy, an expert on currency matters, and a shrewd speculator in stocks.

outline of his theory of colors, obtained his first notion of universal gravitation, and took the first step towards his "method of fluxions" (the differential and integral calculus).* During the next twenty years, however, he largely neglected mechanics for pure mathematics, optics, and chemistry (on which he spent many hours). Although Newton's first published work, in optics (1672), is chiefly remarkable as a contribution to experimental science (pp. 180–181), it is clear that the point at issue between Newton and those of his contemporaries who misunderstood or challenged this first paper was Newton's insistence upon mathematical law, and upon its supremacy over arbitrary mechanistic hypotheses. This was just the point on which Newton also departed from the Cartesians with respect to the system of the universe. Newton held that to every pure color, as made manifest by refraction in the spectrum or otherwise, there belongs a mathematically unique refrangibility. Different transparent materials bend light to varying extents, but (according to Newton and, as it happens, incorrectly) the ratio of bending suffered by the different colors is constant. Color—a physiological sensation—was thus for Newton defined by this mathematical property, to which he ascribed all the phenomena in which colors appear; in the last resort this geometrical individuality of each pure color provided the proof that white light is compounded of them all.

Newton's magnificent later book, *The Mathematical Principles of Natural Philosophy* (1687),† is similarly devoted first to deriving the law of gravitation (the gravitational force between two bodies is proportional to the product of their masses, and to the inverse square of the distance between them), then to proving that this law does explain the phenomena of celestial motion, the tides, and terrestrial gravitation. Because of its universal scope and broad entailments, and

* There is no room in this book for more than allusions to the history of pure mathematics, which has so often been decisively relevant to the history of physics. Solutions of problems of differentiation and integration in special cases had been found by earlier mathematicians; Newton (from 1666) and Leibniz (from 1676) developed general methods of treating such problems, accompanied by a special notation. Leibniz made the first public reference to his method in 1684, Newton three years later. Leibniz's method (from which the modern calculus is directly descended) was quickly adopted by Continental mathematicians. Only the equivalent (but less convenient) Newtonian fluxions were used by English mathematicians till c. 1820.

† As is customary, we shall throughout refer to this as the *Principia*.

because of the majesty of the discoveries in mathematical physics that flowed from it, the discovery of this law has always been regarded as the greatest of scientific achievements. In the eighteenth and nineteenth centuries it was also held to justify, as their culmination, the mechanistic analyses of events that the seventeenth century had gradually elaborated. For the mechanistic view entails the idea that events in nature are lawbound. Descartes, in his *Principles,* had enumerated the "laws of nature." Hence the Creator of the mechanistic universe was also its Lawgiver; as Boyle wrote:

> God, indeed, gave motion to matter. . . . He so guided the motions of the various parts of it, as to contrive them into the world he designed to compose; and established those rules of motion, and that order amongst things corporeal, which we call the laws of nature. Thus, the universe being once framed by God, . . . the mechanical philosophy teaches that the phenomena of the world are physically produced by the mechanical properties of the parts of matter, and that they operate upon one another according to mechanical laws.[6]

Other general propositions that are still generally qualified as laws of nature—the Snell-Descartes sine law of refraction, the law of inertia (first correctly formulated by Descartes) and Boyle's Law, for example—had been enunciated before Newton wrote the *Principia.* None, however, had served as the basis for a mathematically rigorous theory of the type which, coupling the three laws of motion and the law of gravitation, Newton constructed and applied to physics. Each of the earlier laws had been incorporated into, and in part justified by, mechanistic hypotheses. The outstanding distinction of Newton's theory was that it was both mathematical *and* mechanical—a characteristic that all physical theory maintained into the twentieth century.

Possessing this dual character, the *Principia* synthesized the various divergent strands in seventeenth-century mechanics and, by rendering these harmonious, gave the first accurate, mathematical picture of the universe of matter-and-motion. Its three laws of motion derive historically from the cruder notions of Galileo and Descartes; Galileo's law of fall ($s = 1/2\ at^2$) becomes a special case in a general dynamical theory. Descartes had been the first to realize that there must be an inward pull towards the Sun to prevent the planets from moving out of their orbits; he had tried to arrange a pressure in the solar vortex to provide for this. Later, Huygens had investigated (and published without proof in 1673) the law of the force which impels a body to recede from a center about which it revolves. In the *Principia,* gravitational force provides the inward pull, the concept of such a force owing much to the earlier ideas of Kepler, Gilbert, and even Copernicus. Meanwhile, Newton had early arrived at the law of centrifugal force independently of Huygens; from this it followed that if the equal and opposite centripetal force acting on each planet decreased as the square of its distance from the Sun, Kepler's Third Law would be obeyed.* Newton also proved that the same inverse-square law holds with the body moving in an ellipse as well as in a circle, and indeed in any conic section. (Thus the orbit of a comet through the solar system may be a parabola.) Having derived all of Kepler's laws from the dynamics of the inverse-square law, Newton showed that the motions of the planets and their satellites conform to these laws. Further, he gave a dynamical proof that a fluid vortex such as Descartes had imagined could not give them the requisite motions. Newton succeeded in proving an almost perfectly satisfactory theory of the complex revolutions of the Moon, and traced tidal phenomena successfully to the combined action of the lunar and solar gravity; dynamical considerations also enabled him to explain the slight shortening of a clock's pendulum that is required towards the equator and to assert that the Earth's polar axis is slightly shorter than its equatorial axis. The ample confirmation of this prediction in the mid-eighteenth century, together with the return of Halley's comet, as predicted, in 1759, were regarded as the final justifications of the Newtonian system. By this time, indeed, it was universally accepted and had been widely popularized.

A number of important propositions in the *Principia,* particularly that which establishes the concentration of the external gravitational effect of a sphere at its center, depend on the assumption that matter consists of particles which are endowed with gravitation. Newton was an atomist, believing it probable (as he wrote in *Opticks,* 1704), "that God in the Beginning form'd Matter in solid, massy, hard, impenetrable, moveable Particles."[7] Working from the micro-

* In a circular orbit, the centrifugal ($\propto v^2/r$) and centripetal ($\propto 1/r^2$) forces upon the planet are equal, whence v^2r is a constant; which is another way of writing Kepler's Third Law.

physical to the macrocosmic, he showed that on the assumption of such particles, the law of gravitation, and the mathematical theory of motion he had defined, the universe would have exactly the properties of motion that observation discerns in it. As for its other properties, Newton was confident that investigation would reduce these to the same type of explanation. Nature, he wrote, was ever comformable to herself. Thus other phenomena of nature should be reducible to mechanical principles:

> For I am induced by many reasons to suspect that they may all depend upon certain forces by which the particles of bodies, by some causes hitherto unknown, are either mutually impelled towards one another, and cohere in regular figures, or are repelled and recede from one another. These forces being unknown, philosophers have hitherto attempted the search of Nature in vain. . . .[8]

Always supposing that the ultimate causes of phenomena, forces, act between particles (and may not necessarily be capable of direct detection through experiments) Newton has here redefined physics as the mathematical treatment of forces. Partly owing to the difficulties of experimental investigation (Chapter 15) Newton's hopes were hardly realized before the nineteenth century. Then electrostatic and electromagnetic forces were analyzed (by Poisson and Ampère initially; Chapter 17) in ways analogous to Newton's work on gravity in the *Principia*. Indeed, it was shown that these forces too observe the inverse-square law, but the mathematical theory of polarized forces was much more complicated than that of gravity. And, like Newton, mid-nineteenth-century physicists contemplated the possibility of forming a unified theory of forces, towards which the electromagnetic theory of light offered a beginning. At this point, however, the classical, Newtonian picture of a mechanistic universe began to be overwhelmed by its own contradictions.

Consequently, Newtonian physics was always best confirmed by celestial mechanics. Yet in Newton's lifetime the concept of gravitation offered great difficulties. To many the introduction of *force* (i.e., gravity) into the mechanistic philosophy appeared a reversion to occult sympathies. Newton denied that forces could be inherent in matter, yet whence else did they arise? And how could matter exert a force on other matter at a distance unless (as Newton said) by the mediation of something else, through which the force "may be conveyed from one to another"? "But whether this Agent be material or immaterial, I have left to the Consideration of my Readers."[9] None of Newton's readers ever solved this problem. Gravitation remained a fact and a mystery, first demonstrated in the laboratory (as causing a displacement between two suspended bodies) by Henry Cavendish in 1798. Cartesians, supported by all the philosophical adroitness of Leibniz (who was not a Cartesian), continued to frame billiard-ball impact hypotheses until 1750, after which opinion turned steadily towards transferring the mystery of gravity to the "aether," the last home of all mysteries in physics. Newton himself had suggested aethereal hypotheses, though there is little evidence that he believed them. The difficulty did not trouble the mathematicians who were prepared to take forces as given and analyze their effects. The dynamics which Newton had perfected—but to which many others, including Huygens, Leibniz, and Johann Bernoulli, made important independent contributions during Newton's lifetime—was "translated" into the notation of the calculus early in the eighteenth century. In this new mathematical form advances took place rapidly. The important distinction between momentum (mv) and kinetic energy ($1/2\ mv^2$) was clarified. The theory of the motion of fluids, and of the motions of bodies in fluids, which Newton had begun in the *Principia* as another illustration of the fertility of mathematical principles in natural philosophy, was placed on a rigorous basis and widely developed by the Swiss mathematician Leonhard Euler (1707–1783). This theory was of great importance to mathematical physics from the obvious fact that, experientially, bodies do not move *in vacuo*. One aspect of this work was the correction of Newton's marvelously ingenious calculation, from pure dynamical reasoning, of the speed of sound in air. The figure Newton obtained was (as he knew) somewhat too small; a fuller physical analysis traced the discrepancy to the effect of the slight heat created by the motion of sound upon the air transmitting it. This is one example of the way in which the mathematicians of the eighteenth century, with tools rapidly growing more powerful, were able to perfect Newton's picture of the universe. Another is the completion by Clairaut (1713–1765) of Newton's investigation of the lunar motion, and his reinvestigation—leading to a remarkably accurate prediction—of the reappearance of Halley's comet. Always following Newton's path, mathematical physics in the eighteenth century concentrated heavily upon rational and celestial mechanics, seeking to express more rigorously the basic concepts which Newton had first defined and to improve and enlarge the mathematical architecture based upon them. Despite later developments in theory, this architecture remains today in large part valid and useful. Although there were some English contributions, its finished form must be credited to French- and German-speaking scientists, some of them associated with the Academy of Science at St. Petersburg. This was founded by Peter the Great, as part of the westernization of Russia, which entailed the establishment of modern science there.

Newton's universe received its supreme expression in the *Mécanique Céleste* of Pierre Simon de Laplace (1749–1827), published between 1799 and 1825. Laplace's powerful mathematics was avowedly based on the assumption that the future of the universe is completely predictable from its past: given the mass and motion of every particle in it, their future motions (and hence all future events) are in principle calculable. This was the final extension of the principle of predictability, first clearly recognized by the Babylonian astronomers, that had never been wholeheartedly endorsed by Newton himself. For Newton had recognized that the perfection of the universe as God's creation implies its changelessness. Consequently, he thought, the divine providence must constantly forestall those changes in the universe by which natural laws alone would imperceptibly, inexorably, destroy the perfection of the divine plan. (For example, in a finite time the stars must all congregate in one mass under the law of gravitation if they are at finite distances apart, and are repelled by no other force.)* Paradoxically, Newton, who extended dynamics to the heavens, insisted that *as a whole* the universe is static. The second half of the eighteenth century saw the emergence of a completely different view, deriving partly from a universal awareness of the occurrence of change and development that affected biological concepts similarly, partly from astronomical observation.

Tycho Brahe's observation of a nova (1572) suggested that stars might be born and die; Galileo's of sunspots suggested that the brightest star might be filmed over by denser matter. Throughout the seventeenth century Cartesian speculations on stellar evolution were totally unsupported by further evidence. The telescope of this period, while it profoundly modified the aspect of the solar system and so

* For this reason Kant argued (1755) that all stars must rotate about some center in the universe, thus acquiring a centrifugal force.

stimulated notions about life on other planets, was quite unable to add anything new to knowledge of the stars. The size of the universe (two million miles from the Earth to the Sun for Copernicus, eighty million for Newton) rendered them almost inconceivably remote.* Early in the eighteenth century, thanks to the refinement of measurement that the telescope permitted, Halley discerend the proper motions of certain stars; they were not fixed in space. About this time some instrument-makers, believing that the refracting telescope was susceptible of little further improvement, turned to the reflector, first demonstrated by Newton in 1669. Potentially, the reflector was the essential instrument for seeing into space far beyond Saturn, since its large aperture gave it a vast light-gathering capacity. It required an astronomer who possessed both conceptual and mechanical genius to perceive this. Sir William Herschel (1738–1822) constructed one mirror that measured four feet across, although he worked mostly with an instrument of half this aperture. The effect of his observations was to project the chief interest of astronomers outside our little world, and (for the first time) to make the investigation of the cosmos the central task of their science. With him inductively based theories at last replaced ill-founded speculations.

Herschel's first discovery, that of the planet Uranus (1781), was within the solar system. Thereafter he embarked upon a count of the number of stars in the sky from whose apparent distribution (assuming, like Newton, that the stars' real distribution in space is roughly uniform) he concluded that the Sun is somewhat displaced from the center of a lenticular mass of stars (the neighboring galaxy); looking anywhere towards the rim of the galaxy we see the multitudinous stars of the Milky Way—first resolved as such by Galileo. Looking at right angles to the Milky Way, along the axis of the galaxy, we see, as the hypothesis predicts, the stars less densely crowded (Figure 8). Herschel's analysis of his "star gauges" confirmed the earlier hypothesis of Thomas Wright (1750) concerning the nature of the Milky Way, but he went further in postulating, and finding evidence for, a proper motion of the Sun through the galactic stars.

No one was sure, as yet, whether any observable bodies might lie outside the galaxy, or whether this embraced the

* The first really sound calculation of distances within the solar system followed from the observations of the transit of Venus in 1769. Newton had made a close guess of the possible distance of the nearest stars; that was first measured in 1838, by Bessel.

whole universe. Kant, Wright, and Herschel suspected that nebulae did lie outside the galaxy, the last observing these objects with great attention and seeking to distinguish be-between star clusters and gaseous nebulae.* Some of the former he took to be independent galaxies (or "universes," as the word galaxy was not yet in use), and imagined that galaxies in turn had evolved from gaseous collections of matter, represented by the nebulae. The action of gravity would sufficiently account for the coalescence of diffuse matter into stars, in the manner of Immanuel Kant's explanation of the origin of the solar system from cosmic dust (1755).

Laplace, in fact, adapted the idea of the evolution of nebulae to the solar system (1797), supposing that as a gaseous nebula cooled it had slowly contracted; in contracting, its velocity of rotation had increased in accord with dynamical theory, and the forces of rotation had caused a series of rings of matter to break off, somewhat like Saturn's ring. These rings had in turn fragmented, the fragments coalescing into planets whose further contraction and shedding of rings of matter had led to the formation of satellites. Thus the fact that all orbits are nearly in the same plane, and all revolutions in the same direction, would be accounted for. Equally, all other stars might have evolved in a similar way, and some of them possess circumgyrating planets.

The Herschel–Laplace hypothesis of the evolution of the universe, while it did not explain its existence (for matter, as gas, and its motion had to be taken as given) completed the mechanistic view of the universe by extending it back into primordial time. After the fashion of Archimedes, Laplace might have said, "Give me matter and motion and I will construct the universe"—in which, indeed, he would only have been reiterating the ambition of Descartes, a century and a half earlier. The mechanism of force had, however, decisively replaced the mechanism of impact, imposing upon Laplace the assumption abhorrent to Descartes (and Newton) that gravity is inherent in matter. At the same time, too, the attribution of a history to the universe required a distinction between the laws of the universe and its constitution. Newton had recognized such a distinction. But he had supposed providence so to act that the constitution of the universe remained ever the same. Laplace, denying any role in science to God, insisted that as the operation of the laws of the universe continued unbroken its constitution al-

* Herschel's instruments did not permit him to resolve the spiral nebulae (like that in Andromeda) into their component stars.

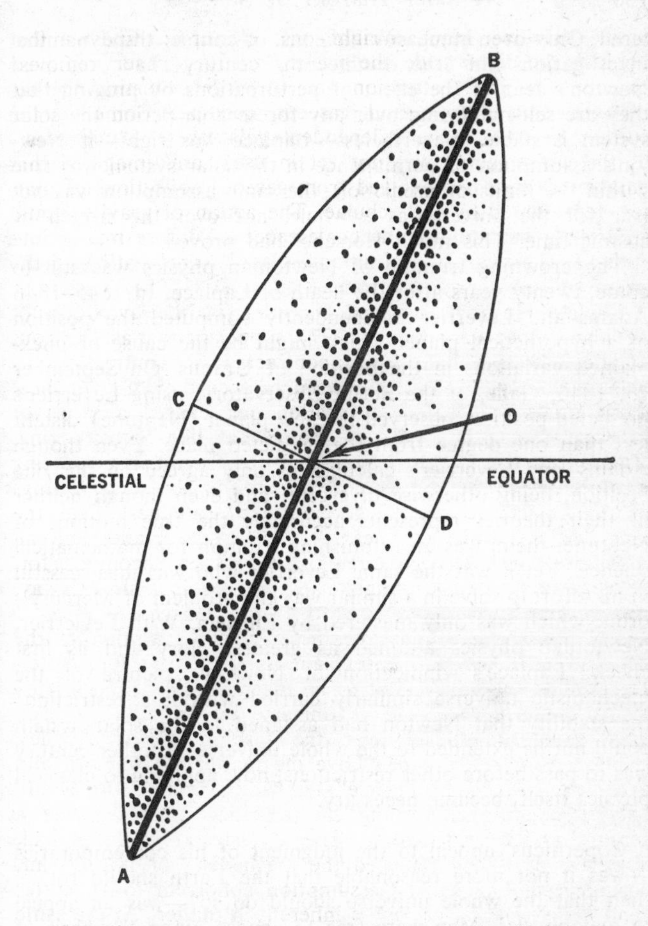

Fig. 8. The Wright-Herschel Hypothesis of the Universe. Suppose that the Universe is shaped like a thick lens, shown in section in the figure, through which the stars are equally dispersed; then an observer near the center at *O* will see many more stars by looking in the direction of the long axis *AO* than along the short axis *CO*. The more densely crowded stars will form a band, *AB*, which corresponds to the Milky Way in the sky; the stars would then disperse gradually on either side of this band, as also happens in the sky.

tered. Only over immeasurable eons, of course: the dynamical investigations of the eighteenth century had removed Newton's fear of the effect of perturbations by proving that they are self-nullifying; over any foreseeable period the solar system is stable. Nevertheless—Laplace was right—if Newton's assumption of permanence in the solar system was true within the limits of prediction, the same assumption was not true for the universe as a whole, nor over the stretch of cosmic time. This much Herschel had proved.

The crowning triumph of Newtonian physics was still to come, twenty years after the death of Laplace. In 1845–1846 Adams and Leverrier independently computed the position of a hypothetical planet which might be the cause of unexplained variations in the position of Uranus. On September 23, 1846, Galle, of the Berlin Observatory, using Leverrier's predicted position, observed the new planet (Neptune) distant less than one degree from the predicted place. Even though Adams' and Leverrier's calculations only agreed on this one position, being otherwise divergent, and even though neither of their theories represents accurately the true motions of Neptune, theirs was an astonishing triumph for mathematical science. Yet it was the same Leverrier who was unsuccessful in his effort to solve in a similar way the problem of Mercury's orbit, which was only mastered by Einstein. With Levierrier, Newtonian physics attained its greatest glory and its first check. Laplace's vindication of Newton's picture of the mechanistic universe similarly carried with it a restriction: the stability that Newton had assigned to the solar system could not be extended to the whole universe. Another century was to pass before other restrictions, now applying to classical physics itself, became necessary.

Copernicus' appeal to the judgment of his contemporaries —was it not more reasonable that the Earth should rotate, than that the whole universe should do so?—was an appeal to metaphysics. Men were free to choose, upon the basis of their preference in metaphysics, the alternative that seemed to them reasonable; they could not choose on the basis of evidence, logic, or any grounds that could be subjected to test. Ultimately very many men adopted Copernican metaphysics and declared that the motion of the Earth was reasonable; only after this did investigation yield some evidence in favor of their conviction. Other men, Galileo and Descartes among them, deliberately rejecting that metaphysic which attributed a real existence in substance or matter to its changeable superficial properties—color, heat, taste, ductility, and so forth—as though these were entities like physical garments that naked matter could take off and put on, concluded that the only realities in the universe were the final terms of their intellectual analyses, matter and motion. Here there is again operative a metaphysical assurance that the macroscopic world familiar from ordinary experience is a sham, a simulacrum, reconstructed by the human organs of perception through their inability to report directly the fine-structure of the true microscopic world, the realm of matter and motion. Hence the new importance of scientific instruments in the seventeenth century, for these seemed to promise knowledge of this realm of real existence.

The mechanistic metaphysic entails the mathematical. Matter can only be understood as magnitude, which is expressed in numbers. Motion can only be understood in terms of direction and speed, equally expressed in numbers. Although the mechanistic picture of nature could usefully be drawn qualitatively, in words, as it was by Descartes and Boyle, the greatest justification for it was the possibility of drawing the picture in numbers. It is because its explanations are quantitative that the explanations of mechanistic science are complete and final, with no logical loopholes; hence their superiority over those of the physics of qualities, Aristotle's physics. When once such mechanistic explanations have been devised, with considerable elaboration and effort, they can be and have been verified by experiment or observation; but at the beginning of things, before the electron had been weighed, before Newton devised celestial mechanics, there was only a metaphysical confidence in what nature is like to assure men that such explanation could be found. As it happens, what has followed from the adoption of Newton's First Law of Motion ("A body acted upon by no forces remains in the same state of uniform rectilinear motion, or at rest") has worked very well. It is useful in our universe and may even be true. It is far more certain that in no universe is there, or could there be, any unique body that satisfies the necessary conditions on which the experiment could be made. Very close approximations to the First Law are, of course, verifiable; but we have only a metaphysical assurance that the simple statement is true.

There is at least one other obvious sense in which the belief that the numerical relationships explored by mathematicians can be applied to events in nature has a metaphysical origin. When a scientist (Kepler, for example, but he was not alone in this) makes it the first axiom of his work that the universe is so arranged that events in it exhibit harmonies or proportions, he is adopting a decided metaphysical position. He may face in either of two directions: he may suppose that harmony in nature is a product of law, hence of mechanism; or he may (like Kepler) be content to derive harmony from confidence that the Creator was a geometer. In either case he acts upon an expectation which is not derived from induction or statistics, but is based upon an a priori confidence that the fabric of the world must be *thus,* and not otherwise.

This, of course, is the meaning of the word "metaphysics" in this context. It denotes the scientist's expectation with respect to the unknown. Today, this is highly conditioned by past experience. In the sixteenth and early seventeenth centuries it was not, for men were endeavoring to free themselves from the experience and authority of the immediate past, though to some extent the new metaphysics of nature that took shape was conditioned historically—by Archimedes and the Greek atomists, for example, and by the mathematical philosophers of the late Middle Ages. Upon these slight anticipations, and in reaction against the Aristotelian metaphysics that had proved so frustrating for science, the natural philosophers of the sixteenth and seventeenth centuries reconstructed their own new expectation of what was to be found in nature. Or, to put this another way, the old terms of explanation having proved as unsatisfactory in general as Ptolemy's epicycles in particular, it was necessary to find new terms upon which alternative explanations could be framed.

The scientific revolution, properly speaking, consisted first of the identification of these terms (matter and motion), then of the discovery of more and more instances of rational connections between them and actual phenomena. The new metaphysic first justified itself, as the new science, in the work of Galileo and Kepler. Its full potentialities were first disclosed in the physical speculations of Descartes. Its triumph was prepared by Huygens, by Hooke, Boyle, and other members of the English school for completion by Newton. By the end of the seventeenth century it was no longer necessary to rely upon a metaphysical expectation of nature, for this had been powerfully confirmed. This did not mean that the mechanistic view of nature was proved, nor did it imply (as the mutual differences between the followers of Descartes, Newton, and Leibniz demonstrate) that its ultimate principles were clear. One is entitled only to say that the natural metaphysics of the scientific revolution had produced a mechanistic, mechanical form of physical theory that corresponded excellently with the known facts. It had yielded a satisfactory picture of a rather limited aspect of nature.

How far could it be extended? Was Newton's treatment of gravity to be, as he hoped, a model for the treatment of all physical phenomena? Would the development of knowledge through experiment permit the introduction of mathematical, mechanical principles into physics and chemistry? These were the problems Newton left to his successors.

Chapter 10

"LET THE EXPERIMENT BE TRIED"

Newton's theory of gravitation, the culmination of seventeenth-century mechanism, was verified by the astronomers' careful observation of the celestial motions. True, the *Principia* contains several experimental demonstrations that the principles of mathematical physics lead to correct conclusions, but Newton's skill as an experimental scientist is best seen in his optical writings. These, rather than the *Principia,* illustrate Newton's declaration that

> As in Mathematicks, so in Natural Philosophy, the Investigation of difficult things by the Method of Analysis, ought ever to precede the method of Composition. This Analysis consists in making Experiments and Observations, and in drawing general Conclusions from them by Induction, and admitting of no Objections against the Conclusions, but such as are taken from Experiments, or other certain Truths. For Hypotheses are not to be regarded in experimental Philosophy. And although the arguing from Experiments and Observations by Induction be no Demonstration of general Conclusions; yet it is the best way of arguing which the Nature of Things admits of. . . .[1]

The distinction between mathematical science (beginning from metaphysically derived axioms and definitions) and experimental science (justifying general conclusions by induction) was one of which Newton was keenly conscious. Yet to suppose that ideas in experimental science were not also profoundly influenced by the prevailing metaphysics of the seventeenth century would be a grave mistake.

The tradition of logic stemming from Ockham had emphasized the making of experiments as a method of proof. If some property is attributed to nature, then experiment should reveal it. The emphasis of the seventeenth century was upon experiment as a method of fact-finding, a stage preceding the formation of general ideas which in turn, of course, should be verified by fresh experiments. This version of empiricism is very ancient, since it is an elaboration of all trial-and-error procedures that yield formulas and recipes of the cookbook type. There could have been no practical knowledge of chemistry without such empiricism, nor of magnetism. In the same sense every physician was an experimentalist in so far as he varied drugs and treatment to obtain the best effect upon his patients. In the later sixteenth century especially, as medical opinion inclined to place chemical remedies alongside "Galenicals," or even to rate them superior to the latter, there was a free empiricism in the treatment of disease. Similarly, Europeans experimented with the products of the New World, not only drugs like guaiacum, but corn, potatoes, chocolate, and tobacco. Most of all, perhaps, experiment implied in the sixteenth century exploration of occult things in nature, the mystery associated with the vacuum, magnetism, optical illusions, secret communication, and alchemy, not to mention esoteric processes and devices of all kinds. While such exploration often ended in charlatanry and conjuring tricks, it also produced some useful results. Baptista Porta's *Natural Magic* (1588) is the classic example of the farrago of sense and nonsense that experiment denoted.

It was the principal philosophic endeavor of Francis Bacon (1561–1626) to rescue experiment from this dubious context, and to render it the indisputable basis for scientific explanation. A fierce critic of scholasticism, which he thought taught men only to spin fine cobwebs of wordy speculation, and of Aristotelian philosophy, which had committed the error of jumping at once to general principles without demonstrating their truth by induction from experiments, Bacon held that the first task before science was to dismiss all theories and to compile "histories" of nature (factual encyclopedias concerning all manner of phenomena). Only when all the facts were known, as they might be by systematically making all observations and experiments that suggested themselves, would it be worthwhile to attempt to frame general ideas. This task could be reduced almost to rule by sifting from the collections of facts the characteristics common to all instances of a single phenomenon, noting the apparent exceptions, and so forth. Bacon was aware that this sifting of the factual material might suggest further experiments to strengthen or confute an hypothesis, but for the most part he supposed that adequate theories could be obtained by generalizing immediately from the facts. Bacon attached little importance to the mathematical development of ideas, or to scientific imagination, or indeed to reason except as directed by a strict application of induction.

After two centuries of praise as the "legislator of modern science," Bacon's reputation has now fallen to an undeservedly low level. It is true that he understood nothing of mathematical science, being far more interested in natural history. It is true also that he ignored or rejected the original work of his immediate predecessors (Copernicus among them) and contemporaries, one of whom (William Harvey) recorded caustically that Bacon "wrote science like a Lord Chancellor." Bacon's attempt to provide a logic of scientific discovery has been derided; no one, it has been said, made a discovery by Bacon's method—not even Bacon himself. Finally it is clear that Bacon was carried away by (and also, being a pioneer, helped to determine) the metaphysic of his age: he was a mechanical philosopher long before Galileo or Descartes had declared themselves to be such. As a demonstration by induction from empirical facts, for instance, Bacon's famous example of the use of his method to derive a general notion of heat is quite worthless; as a revelation of Bacon's metaphysic, and of the arguments that could be used to defend it, these pages are invaluable.

In one aspect of his philosophy, then, Bacon determined Robert Boyle's ambition to replace Aristotelian qualitative explanations in physics by empirically justified mechanistic explanations. Boyle and his colleagues in the Royal Society, half a century after Bacon's death, looked back to him with veneration, regarding their own work as implementing his philosophy. To them, indeed, he was an inspiration if not a model—for Boyle's or Newton's experimental science was very different from Bacon's. The Royal Society's motto *Nullius in Verba* ("On the word of no man") echoed his reply to any report or speculation: *Fiat experimentum* ("Let the experiment be tried"). It is false to suppose that science has developed only by brilliant theorization, mathematical or not: modern science is the product of endless patient labor devoted to the acquisition of facts for their own sake. In science (as in everything else) the moments of intellectual excitement are only earned by much dull plodding. Bacon was perfectly right to insist upon this, especially with regard to the biological sciences towards which his own thoughts turned. He was right, too, in declaring that the new metaphysics of nature needed empirical confirmation. What he did not accurately foresee was the way in which experiment in science would be guided by its theoretical outline, which is very different from a haphazard "gathering of instances" (though he did not wholly favor that either). Nor did he foresee how the experimental lines of an investigation might be shaped by a daring projection of an idea forward from the existing state of knowledge in order to demonstrate that "this is how things must be." Bacon was the all-time intellectual democrat. Although a contemporary of Galileo, Kepler, and Harvey he never enjoyed the faintest notion of what it is for one individual to transform a science.

Hence he laid stress on collective effort. The complexity of nature was just too great for one man to master. Here also Bacon spoke sensibly; he foresaw the utility of institutes for research in science. The first of the national scientific societies, the Royal Society for the Promotion of Natural Knowledge (London, 1662) and the Académie Royale des Sciences (Paris, 1666), admitted their indebtedness to Bacon's sketch in the *New Atlantis* (1628). Earlier groups or clubs of men interested in science or medicine had either been highly informal (in some, the members were scattered through many towns, so that they could never meet together) or, like the Accademia del Cimento in Florence (1657–1667; it carried out a valuable program of experiments in physical science), were short-lived creations of private patronage. The important notion of the later societies (Baconian in origin) was that they should serve both as a tribunal to approve new work in science or reveal its defects, and as a means of disseminating what seemed sound and new.

Such schemes, in so far as they envisaged planned and cooperative research, were far too ambitious. The finance and the facilities for them did not exist. In the end experimental science flourished through the efforts of individuals, and in the directions along which it was led by individual interests. For example, one whole branch of experimental science in the seventeenth century, pneumatics, sprang from

Galileo's discussion of the vacuum in his *Discourses* of 1638.* Was the force with which nature resists the formation of a vacuum limited, as he suggested and as the failure of suction to raise water more than 32 feet seemed to prove? Experiments at Rome with water, and later with mercury in Florence, demonstrated the appearance of an empty (but light-transmitting) space when the fluid was allowed to descend to a stable level in a tube more than 32 feet (or 30 inches) high. The mercury experiment of Evangelista Torricelli (1608–1647) confirmed his hypothesis that it was not nature's intolerance of a vacuum but the weight of the atmosphere that kept the fluid suspended: for the weights of water and mercury (per unit area) were equal. In other words, the column of fluid was upheld by a determinate pressure, which Torricelli and others surmised was caused by the weight of the air. (There was nothing new in the idea following from this, that the atmosphere has a limited height.)

Could this surmise be rendered firmer? Several philosophers reasoned that if the column of fluid in the barometer (as it was now called) was supported by the air, it should be shorter at high altitudes than at low altitudes. The experiment was suggested by Blaise Pascal (1623–1662) to his brother-in-law, who carried a barometer up the Puy-de-Dôme, a mountain in central France, and discovered that the atmospheric pressure was several inches less (as measured by the length of the mercury-column) at the summit than at the base. This was hailed as a crucial experiment, and a triumph for mechanistic physics; Pascal went on to write a little book on pressure in fluids (both elastic like air and incompressible like water) in which he showed with the aid of many experiments that pressure was transmitted equally in all directions; he also distinguished between pressure and weight. The atmosphere does not weigh down on a barometer placed in a sealed vessel, but the atmospheric pressure is maintained within it.†

Pascal's experiments had very obviously not all been performed, since some of them required the experimenter to sit at the bottom of a tank of water many feet deep. Like Galileo and many others Pascal used the "thought experiment" as a method of reasoning or of winning conviction. Others, however, were making true experiments, especially

* Of course, this was a traditional problem, and the experimental approach (the force required to pull the piston down a closed syringe, etc.) well known.

† Descartes had urged that in such a case the mercury would descend.

those who were trying to investigate the "emptiness" of the (partial) vacuum. It was found that magnetic force, as well as light, traversed it, while the passage of sound through a vacuum was fairly soon found to be caused by faulty experimentation. Many, including Descartes, were most reluctant to believe that any space could be empty of matter; if it were quite, or even largely, empty of the finest perceptible substance (air), the space might still be full of such an exceptionally fine matter, or aether, as could pass through the pores of all solid bodies. This aether would then be the vehicle of light and magnetism.

Naturally, experiments upon this undetectable matter were impossible. It was nevertheless useful to explore the properties of the apparent vacuum; Torricelli had prepared an experimental chamber by widening out the top of the barometer tube. When the tube was filled with mercury, then inverted so that the fluid ran down, a sizable space was left. This method was continued by the Accademia del Cimento. Meanwhile at Magdeburg in Germany the learned mayor, Otto von Guericke (1602–1686), contrived to exhaust vessels by means of a large syringe or pump, his chief object being to demonstrate the phenomenal compression exerted by the atmosphere on large surfaces. News of his experiments reached Robert Boyle at Oxford in 1657 or 1658. His eagerness to construct a true air pump, with which laboratory experiments could be made, was satisfied by his ingenious assistant, Robert Hooke (1635–1703). The experimental vessel in Boyle's pump—the first piece of modern physical apparatus—was fitted with a removable plate for the insertion of the subject of the experiment, and with a key so that simple movements could be effected within after the air was exhausted. Boyle's first series of experiments on the "spring and weight of the air" (published in 1660) confirmed its elasticity in a variety of ways, and the doctrine of atmospheric pressure; he went on to examine thoroughly the cessation of respiration and combustion that occurred *in vacuo*. It was in a defense of this book against criticisms that Boyle enunciated in 1662, from a table of relative pressures and volumes, the eponymous Law, $pv = k$. At subsequent periods he published two further accounts of pneumatic experiments, but these are less significant than the first.

Through Boyle the air pump became both a standard piece of scientific equipment and an eloquent exponent of the mechanistic philosophy. At the same time his experiments with it, and those of others of different types, brought fresh interest to the old analogy between breathing and burning to which a group of English scientists gave a new, mechanistic form. Thus the first modern attempt at biochemistry carried a strong flavor of physics. Burning, according to the mechanical philosophers, was simply the disappearance of the particles of a hot body into the atmosphere, and flame a hot, shining vapor. The action of the air was clearly necessary to this dissolution of the combustible, for it ceased when the air was removed: probably the air played a similar role in animal physiology, whose processes likewise generate heat and cease if air is denied them. Robert Hooke, in 1665, drawing the obvious chemical parallel, said that air was the solvent of bodies, but not all of the air. It was well known that combustion and respiration stop long before all the air in a closed vessel has been "used up"; this point was well illustrated in experiments of John Mayow (1645–1679), who placed in a jar inverted into water a mouse, or a candle, or both (Figure 9). The air appeared to be partially "used up,"

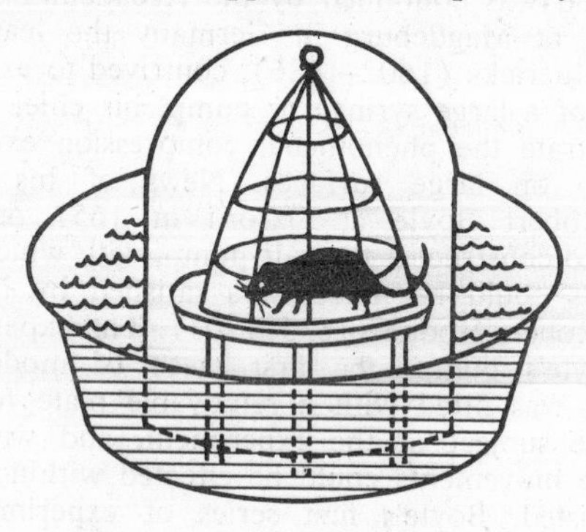

Fig. 9. Mayow's apparatus for experiments on respiration. A mouse, secured by a cage, is placed under a glass jar. The air in the jar is sealed by immersing it in a tub of water. A small pipe is used to bring the water to the same level inside and outside the jar at the start of the experiment. As the mouse breathes, the water inside the jar rises.

since the water rose inside the jar, but after a little while the flame went out and the rise of the water ended. Such experiments suggested that something in the air was involved in respiration and combustion, while the rest of the atmosphere was inert. Hooke, Mayow, and the physiologist Richard Lower (p. 202) agreed in qualifying this constituent in (rather than of) the air as "nitrous," or as consisting of nitro-aerial particles. Such particles, they held, removed the combustible or "sulphureous" * particles from that which burnt.

Why "nitrous"? Because the action of air in damp places, especially where animal excrements are present, causes the formation of niter (or saltpeter); because niter supports combustion where the air is excluded, forming, in gunpowder, a powerful explosive; and because niter—or similar salts—acts as a fertilizer (like dung) to promote the growth of plants. These rather devious analogies had suggested to a Polish alchemist named Michael Sendivogius, and others who adopted his ideas through the first half of the seventeenth century, that niter is a "food of life," like air, and hence (completing the circle) of flame. Boyle, for instance, suggested that something of the nitrous kind might restore the vitiated air of a closed space (he was, of course, wholly unaware that niter is a good oxidizing agent). Boyle, too, had shown that niter is itself a compound substance. In the terms of the mechanistic philosophy, it seemed reasonable to suppose that a certain type of nitrous particle existed in the air, which became "fixed" with other particles to form the solid niter, and that these nitrous particles, in their reaction with sulphureous ones, were responsible for the heat-producing phenomena of combustion and animal physiology. Certainly nothing that was discovered in the experiments made between 1660 and 1680 conflicted with this theory, and in so far as they proved that a *part* of the air was essential to these phenomena the experiments seemed to confirm it.

These studies furnish an excellent example of the interrelation of theory and experiment. Those who made them were certainly not proceeding in the detached, inductive manner prescribed by Bacon, for they clearly produced the experiments as evidence for the theory, not the experiments as data from which the theory might be induced. Boyle, the most cautious of this group in theorization, rejected the ni-

* The term "sulphureous" derived from the alchemical theory that there are three immediate constituents of bodies, or principles: salt, sulphur, and mercury. These were not identified with the common substances of the same names. *Salt* was typified by solubility, incombustibility, and solidity; *sulphur* by oiliness and combustibility; *mercury* by fluidity and metallicity.

trous theory on the ground that nothing nitrous can be detected in the air. Mayow, the wildest speculator, sought to make nitro-aerial particles responsible for half the phenomena of nature. All of those involved did good experimental work, however, some of it physical, some of it chemical, some of it physiological, and tried to tie the disparate fragments of evidence thus obtained into this single consistent theory. And their speculations, though hindsight may render them extravagant, at least took the line of explaining many phenomena by applying to their different circumstances a single fundamental idea; one mechanical hypothesis which might work out in several different ways. Given the metaphysical principle that everything is to be reduced in the last analysis to the motions of particles, their explanation of their experiments was rational; equally, however, it is clear that this whole experimental development is conditioned by the experimenters' adoption of this principle, an adoption that follows Bacon's practice rather than his precepts.

Another important, though less flamboyant, line of seventeenth-century experiment also stems from Galileo, though in this case its affiliations were rather with his mathematical than with his physical ideas. If the universe is mathematical as Galileo maintained, and more particularly if his kinematics is correct, it should be possible to find verification in experiments. He had himself proposed the equal swings of a pendulum as an empirical justification of his idea of inertia, and had experimented on the times taken by a ball to roll different distances on inclined planes in order to confirm his law of fall. Many other experiments to test his kinematics were possible: on projectiles, on falling bodies, on the pendulum, on jets of water; a number of these were performed by the members of the Accademia del Cimento, and by other scientists. Inevitably, it proved difficult to obtain direct results exactly in accord with Galileo's theory, since that theory is only applicable to a vacuous, Euclidean universe. Later, more attention was paid to the physical realities; Huygens and Newton, for instance, both experimented upon, and developed mathematical theories of, the resistance offered to motion by air and water. In the eighteenth century mechanical friction was similarly subjected to experiment in the search for mathematical laws. Descartes' emphasis on impact phenomena likewise prompted both experimental and theoretical inquiry into the laws governing the partition of motion between colliding bodies.

All such experimentation in mechanics (except that on

friction) was of course distinguished from other experiment in physics and chemistry by its combination with a sophisticated mathematical attack. Mechanics was a mathematical, not an empirical, science, in which the chief function of an experiment was to determine the actual value of some physical constant—like *g*—required for the application of theory. In the eighteenth century this theory was often explained and illustrated with the aid of experimental demonstrations by popular lecturers, for the benefit of nonmathematical audiences, but no important discovery in mechanics was ever deduced from an experiment. It was quite otherwise in electricity, magnetism, acoustics, and optics. While the two first-named branches of physics proved unfailing stimuli to curiosity, they made no further progress after Gilbert's classic book (1600), except that von Guericke contrived the first simple electrical machine. Acoustics made some not very remarkable progress: harmonics were better understood and pitch was definitely related to frequency of vibration. But the truly empirical physics of the seventeenth century is to be found in optics.

Until about 1660 the main interest was in geometrical optics, stimulated by the use of lenses in the telescope and microscope. Following upon Kepler (1611), Descartes, with the aid of the sine law of refraction (which he was the first to publish) gave a good account of image formation by a lens (1637) and improved the explanation of the rainbow. Imperfect lenses called attention to the colors formed on refraction, for the colors were seen as fringes surrounding the image in the instrument. Partly for this reason, and partly because Descartes had attempted to explain light and color along with everything else in his system, physical optics became a center of activity after 1660.

Here again it is difficult to dissociate theory and experiment. Two important new phenomena of light were described in 1665; diffraction (discovered independently by Robert Hooke and Francesco-Maria Grimaldi), and the colors of thin plates (independently discovered by Hooke and Newton). A third, double refraction, was discovered in 1669 by Erasmus Bartholinus. Each of these phenomena suggested to one investigator or another that light must consist of a pulse-like motion, or longitudinal wave, through the aether. Such a view, deriving from Descartes' hypothesis that light is a pressure in the aether, exploits an analogy between light and sound. It was strongly pressed by Hooke and Christiaan Huygens. The latter, in his *Treatise of Light* many years later (1690), laid the mathematical foundations of wave theory.

But wave theory suggested no obvious physical explanation of colors. (Newton, who rejected the wave theory, first pointed out that color, like pitch in sound, might be related to frequency; this would have agreed with his own doctrine that all colors are confused in white light, but the wave theorists did not thank him for the hypothesis.) Hooke, however, proposed in his book on the microscope (*Micrographia,* 1665) that the pulses of light might be bent on refraction, and that these bent pulses might cause the sensation of color. He showed how a similar bending might be produced by the interaction of refraction and reflection at the surfaces of thin plates, giving rise to different colors as the interval between the surfaces of the plates varied. To Hooke, then, colored light was a modified form of white light.

Hard upon these discoveries and speculations came Newton's paper on light, published in 1672. He had already established his mastery of practical optics by making the first useful reflecting telescope, which he had presented to the Royal Society in London. Newton had begun his optical experiments with a familiar but undervalued instrument, a glass prism. Some early observations taught him that colors are differentially refracted: a straight line, part red, part blue, no longer appeared straight when viewed through the prism. When he projected a narrow beam of light through the prism —with typical forethought, he cast the spectrum upon a distant screen, so that it was large and perfect—he found that the spectrum was not round, like the beam, but elongated:

> Comparing the length of this colored *Spectrum* with its breadth, I found it about five times greater; a disproportion so extravagant that it excited me to a more than ordinary curiosity of examining, from whence it might proceed.[2]

Analysis, supported by further experiments (not all of them revealed by Newton in his first paper) persuaded him that this stretching of the spectrum could only occur because, again, colors were differentially refracted, violet always a constant amount more than red (Figure 10). And this effect seemed to be independent of the glass used; hence lenses must always yield colored images, and the perfection of refracting telescopes be impossible. As already mentioned, this experimental investigation convinced Newton that what we

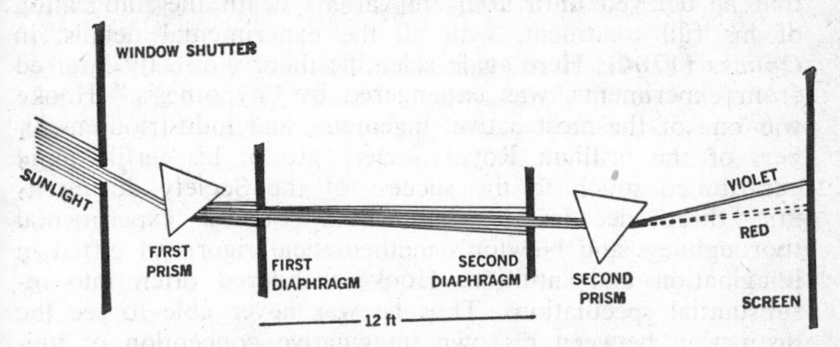

Fig. 10. Newton's "Crucial Experiment" on refraction. The second prism receives a narrow beam of colored light from the first, through two diaphragms. By rotating the first prism, the second may be made to receive any spectral color. The position of the image on the screen can be shown to vary with the color, although all the colored rays received by the second prism are parallel.

perceive as color is a mathematically definite property of light.

While Newton was correct in principle—for we now identify the frequency or wavelength of each color as a mathematically defined property—he was mistaken in supposing that dispersion, which causes the elongation of the spectrum, is a constant function of refraction. Even at that time this was doubted, and in 1733 Chester Moor Hall constructed an achromatic telescope. The theory of achromatism, by which lenses of different kinds of glass having dissimilar dispersion/refraction ratios are combined so that the red and blue images form in the same plane, was further investigated by Euler and the London instrument-maker John Dollond, who began to manufacture corrected lenses from 1759. Improvement in lens theory and optical glass manufacture made possible the construction of refracting telescopes up to about 36 inches aperture by the mid-nineteenth century; this is still nearly the useful limit for such instruments. Similarly, when the method of correction was extended to microscope objectives a great increase in magnification and (more important) resolution of fine-structure became practicable.

Newton's assertion that the primary colors in the spectrum are simple and white light compound was immediately challenged, especially by the wave theorists. In the controversy that followed Newton conceived such animosity to Hooke that he delayed until after the latter's death the publication of his full treatment, with all the experimental details, in *Opticks* (1704). Here again scientific theory, directly inferred from experiments, was endangered by "hypothesis." Hooke was one of the most active, ingenious, and industrious members of the brilliant Royal Society group; his fertile mind contributed much to the success of the Society during its first three decades. Lacking both Boyle's experimental thoroughness and Newton's mathematical rigor, yet gifted in imagination and intuition, Hooke wandered often into insubstantial speculations. Thus he was never able to see the distinction between his own imaginative conception of universal gravitation (developed over the years 1660–1680) and Newton's demonstration of the law of gravity as an essential feature of the mathematical-mechanical physics of nature.

No more than Hooke and other wave theorists did Newton declare the actual physical nature of the colored constituents of white light. He possibly believed that each pure colored ray consisted of a stream of minute particles, differing (according to the color) in either mass or velocity; it would be possible also to examine how such a confused mass of particles might be separated mechanically by a refracting substance, somewhat in the manner of a modern mass-spectrometer, to form the distinct but intermingled colors of the spectrum. Only by supposing light to be corpuscular, an emission of matter from the source, could its rectilinear propagation be explained; wave motion (Newton argued) always bends into the shadow.* Sound, for instance, does not travel along straight lines only. At the same time Newton's own study of the colors exhibited when light passes through thin plates compelled him to suppose that the traveling ray is characterized by certain regularly periodic states ("fits") which represent, in a sense, a wave motion superimposed upon the stream of particles. Meticulous quantitative experiment enabled Newton to compute the intervals between these "fits" in the various pure colors, intervals which correspond closely to modern values for the respective wavelengths.

The statement that Newton taught a corpuscular theory of light is false, though it was accepted as true through much of the eighteenth century. Like the twentieth-century theory of light, Newton's admitted both a corpuscular and an un-

* Huygens countered this argument by showing mathematically that the amount of "bending into the shadow" is very slight if the wavelength is very minute; in the early nineteenth century this bending was identified with diffraction.

dulatory aspect, although the resemblance is of a very loose character. This was clearly brought out by Thomas Young (1773–1829) who began the reinstatement of the wave theory in 1801. In that year he formulated the concept of optical interference to explain how thin-plate and diffraction colors might arise from the combination of wave motions traveling along slightly different paths. In his discussion of wave theory Young borrowed much from Newton, though in his view Newton's addition of a corpuscular concept was needless. As the wave theory developed in the early nineteenth century, especially through the work of Fresnel (pp. 259–259) it departed further from its likeness to Newton's hypothesis of "fits"; nevertheless, the effect was to emphasize one part of Newton's concept of light while abandoning the other. Still later investigations—among them the discovery of the light-pressure predictable from Newton's corpuscular hypothesis—in turn necessitated reversion to a particle concept, that of the photon.

To the nineteenth century wave motion implied the existence of an aether in which that motion occurs, "the subject of the verb, *to undulate.*" It now seemed that the passage of light, as well as the phenomena of gravity, magnetism, and electricity, required this mysterious entity for its explanation. Newton was held to have blessed these later speculations. Curiously enough, the very physicists of the eighteenth century who had been most eager to fasten the corpuscular idea of light upon Newton had also been most convinced that he believed the ultimate resource of explanation in physics to lie in the aether. (The combined attribution is dubious as well as inconsistent.) It is certain that, when Newton spoke of what he knew, he said no more of the aether than he did of light particles—he confessed that he knew not the cause of gravity. In some letters, however, and above all in the *Quaeries* appended to successive editions of *Opticks,* Newton allowed himself freedom to speculate about what he merely guessed. Among Newton's guesses about the nature of things (some of them mutually contradictory) he did indeed suggest a possible mechanical cause, or aether; partly he did this to refute the charge that the law of gravity was unintelligible, because he had offered no mechanical hypothesis to account for it. By a very odd defect of logic, some of the later scientists who quoted Newton's own dicta against the confusion of mere speculation with scientific theory properly speaking, also argued that these Newtonian speculations were as good as proved.

While the development of mathematical physics was transferred to the Continent, a generation of British scientists* preferred *Opticks* (as they understood it) to the *Principia* (which they understood very little). They developed a picture of Newtonian science in which the master appeared as a Baconian empiricist, eager to find the ultimate mechanical explanations of things in an aether. In this image Newton gave a renewed impulse to experimental science in the eighteenth century, encouraging men to believe that if some problems in physics could not yet be analyzed mathematically, nevertheless it was proper to explore them by experiments and seek to resolve them by mechanical hypotheses. If Newton had explained gravity by the aether (!), they could attempt the same for electric attraction. Though any connection between eighteenth-century theories of impalpable fluids (electricity, heat, phlogiston) and Newtonian mathematical physics may seem remote, it was furnished (inadvertently perhaps) by Newton himself, and the connection was not uncreative.

This version of Newton's physics as empirical and inductive authorized attempts to demonstrate mechanics experimentally (pp. 178–179); more interestingly, it encouraged experiments designed to develop hints Newton had let fall in his *Quaeries*. Newton, like every seventeenth-century physicist, had been interested in the phenomena of heat and the perplexities of defining adequate concepts for expressing them. An important, if humdrum, step was the definition of standard, reproducible thermometric scales by G. D. Fahrenheit (1686–1736), R. A. F. de Réaumur (1683–1757), and Andreas Celsius (1701–1744). Benjamin Franklin and others, like Boyle and Newton before them, studied the differential absorption of solar heat according to color. Much later (1800), examining the heating effect of different portions of the solar spectrum, Sir William Herschel discovered that it extended beyond the visible range into the "infrared." Ideas of the radiation of heat were still rudimentary, however; it was not long since the similar radiation of cold had been debated. Yet Herschel's discovery cast doubt on Newton's guess that rise in temperature was a manifestation of the action of light upon matter. By this time the universal seventeenth-century idea of heat as a "mode of motion," a mechanical agitation of the minute particles physically ap-

* This includes (of course) some American figures, e.g., Benjamin Franklin (1706–1790). There were, besides, a few Continental empiricists who belonged to this British school.

parent in the expansion of heated bodies, had been supplemented by an "aetherial" hypothesis. On this view the "matter of heat" ("caloric" to Lavoisier and later writers) was an impalpable, weightless fluid, flowing down a gradient from hot to cold; this fluid was apt to motion and to move the particles of bodies, as the heat produced in friction demonstrated. The hypothesis thus explained conduction, but not radiation. As the quantity of this matter or caloric in a body increased, a thermometer attached to it would show a rise in temperature. At first no distinction was made between the quantity of heat that flowed into a cooler body from a source, and the ensuing increase of temperature, since the two seemed to be proportional. Joseph Black (1728–1799) recognized about 1760 that in similar circumstances the variation in temperature between two different substances having equal volume was dependent, not on their density, but upon a constant for the substance, its specific heat. He found, for example, that the heat capacities of mercury and water were almost as 1 to 2, although mercury was thirteen times more dense. Hence it was necessary to distinguish between temperature (measured by the thermometer) and quantity of heat (today measured in calories), which could be conveniently compared by experiments on the quantities of ice at 0°C melted before different substances were cooled to 0°C.

The concept of specific heat was perfectly in accord with the fluid theory, and the notion of this constant was no more troublesome than that of specific gravity already known or those, like specific resistance, discovered later. Black's other discovery about heat was more fundamentally puzzling, since it implied a gap in the continuity of phenomena. He pointed out that ice, in thawing at 0°C, absorbed heat while changing from solid to liquid without altering its temperature; similarly water vapor at 100°C gave out heat on condensing into liquid without change of temperature. The heat involved in such changes of state was latent, revealed in qualitative changes but not by the thermometer. Theories are often surprisingly elastic; the idea of latent heat was accepted into the late eighteenth-century definition of substance as matter plus caloric without qualm, though in fact the discrepancy between (a) increase of caloric causes increase of temperature, and (b) increase of caloric causes only a qualitative change, was to prove of fundamental significance to ideas about the constitution of matter.

Matter could not consist merely of minute particles bathed

by a soup composed of the aether and the heat-fluid; it was necessary to add an electric fluid * as a further ingredient. Early in the eighteenth century the "electric fire"—so-called for obvious reasons—was indeed tentatively linked either with the matter of heat, or the aether, but as fresh information about conduction and so forth had to be fitted into hypotheses of electricity, the electric fluid was inevitably more sharply characterized as a distinct entity, weightless and impalpable like the matter of heat. The postulation of this entity, by which Franklin was enabled (1751) to account for a wide range of empirical observations, depended in the first place on the manifestation of its effects. Electricity was the first branch of modern physics to owe its origins wholly to experiment, and nothing to the development of a chain of ideas stretching back to antiquity. In this sense—though not in the accompanying oscillations of speculation—electrical science was truly Baconian. Not less, to readers familiar with the enigmatic allusions made by Newton at the end of the final version of the *Principia,* was it Newtonian.

Only the so-called "static" or frictional form of electricity was known until the close of the eighteenth century. Its most striking manifestations required large machines in which the charge was collected from plates or cylinders of glass rotating against leather pads. These could be even more spectacular than the air pump, prompting large circles of monks (or soldiers) to leap convulsively into the air, yielding foot-long sparks, and killing small animals. For serious research the charge on a smartly polished glass rod was equally useful. An electrician of Franklin's day, however, knew that any substance could be electrified or charged, even those that conducted electricity (preeminently, the metals) if they were insulated while being rubbed, as also that there were two "forms" of electricity (corresponding to the modern positive and negative charges) as there were two magnetic poles, like repelling like and unlike forms attracting each other. He was aware that a charge could appear at the end of a long length of wire and that it could be built up in the plates of a Leyden jar (condenser) to yield a formidable shock. He was thus prepared to entertain an idea of electricity that embraced both the notion of fluid flow also associated with

* No doubt a "unified fluid" theory would have reduced all these to the aether, but in fact eighteenth-century scientific explanations all turned on the postulation of specific fluids for the various branches of physics.

heat, and the notion of attraction familiar from gravitational theory and magnetic experiments.

The distinction of Franklin's electrical theory was that it synthesized the Newtonian concept of an attractive force between the ultimate particles of matter with the postulation of a single electric fluid, consisting of mutually *repulsive* particles. Thus:

> Electrical matter differs from common matter in this, that the part[icle]s of the latter mutually attract, those of the former mutually repel each other. . . . But though the particles of electrical matter do repel each other, they are strongly attracted by all other matter.[3]

In Franklin's view positive charging consisted of pumping into a body, from the great reservoir in the Earth, a greater quantity of the electric fluid than it normally contained; conversely, negative charging involved a forced withdrawal of some of the body's normal electricity. Hence no body could be charged in complete isolation. In like manner Franklin explained the action of the Leyden jar; it could not accept a charge on one plate unless there were an equal deficit on the other, so that its net gain of electricity was zero. In fact, the only phenomenon Franklin's theory could not explain was the repulsion of two negatively charged bodies, a point of which he was at first unaware. For, since negative charge indicated a deficit of the electric fluid, the normal matter in the two bodies should have evinced a slightly greater mutual *attraction* between them than usual.

Franklin's experiments, and still more his comprehensive theory, gave him a distinguished reputation among men of science in Europe; to a far wider public he was the hero from the New World who had tamed lightning. Franklin's proposal of a lightning-conductor experiment to test the hypothesis (it was not new) that the lightning discharge was electrical, successfully practiced in France, swept the imagination of Europe.* Here at last was a truly majestic demonstration both of human intellectual power displayed in science, and of its practical utility in calling Nature to heel. Franklin was by no means the first North American to enjoy scientific

* Franklin (and all liberals) supported the use of sharp-pointed conductors to lead the electric fire from the air harmlessly into the ground. George III and all enemies of freedom believed in blunt-ended conductors, to discourage the lightning-stroke. For reasons of politics and emotion the pro-Franklin enthusiasm was wilder in France than in England.

repute in Europe (which he had visited as a young man before Newton's death, and long before his own triumphs there in many capacities). European naturalists were already grateful to their colleagues across the Atlantic for their skill and pains in collecting new specimens, perpetuated in such species as *Gardenia*. And much earlier still John Winthrop, first Governor of Connecticut, had been a highly esteemed Fellow of the then infant Royal society. Yet Franklin, as a physicist (though far from a mathematician), was the first American to come close to Newton, as he was assuredly also the first man of the people to be lionized in society as a hero of science.

With Franklin's robust, striking personality—clearly reflected even in his scientific writings—the first critical phase of experimentation on electricity ended. True, there were far more elegant and exact electrical experiments still to be performed in the eighteenth century, such as Coulomb's establishment of the inverse-square law of electrical attraction (1784–1785) * and the delicate, far-reaching work of Henry Cavendish (1731–1810), hidden in his notebooks, which in many respects anticipated the later investigations of Michael Faraday (p. 253). Cavendish, like Joseph Priestley (p. 194), investigated the flow of charge along conductors and the resistance to it offered by different materials. But so far, despite the obvious mechanical analogy expressed in the idea of an electric *fluid*, the continuous flow or current of electricity was quite unknown. Apart from the spark, in which its fiery nature was revealed, electricity was best known by the mechanical effects of attraction and repulsion; further, it was mechanically produced, by friction. Although Franklin's theory postulated that the electric fluid was universally associated with matter, it became manifest only when its normal state was disturbed. Thus it was indeed surprising when at the end of the century electricity was detected as a calm current, not a violent and rapid discharge; as a product of normal chemical processes, not violent mechanical actions; and as totally devoid of obvious mechanical or qualitative effects.

The discovery was made by accident, and in a manner that ensured its initial misinterpretation. About 1780 Luigi Galvani (1737–1798), an anatomist working at Bologna, learned that if a scalpel were pressed against the crural nerve of a dissected frog's leg while sparks were emitted from an

* With the same apparatus, the torsion balance, Coulomb also demonstrated that the same law holds for magnetic force.

adjacent electrical machine, the muscles of the leg were violently contracted. (Galvani had in fact observed electromagnetic induction, but of this he was totally unaware.) Galvani took the passage of electricity through the frog's leg for granted, since it was well known that muscular contractions occurred when discharges were passed through dead animals. Later, he noticed that lightning flashes could have the same effect. Then, again quite by accident, he found that if two different metals were in contact, the one also touching a nerve, the other also the muscle, contractions took place. They occurred also when a single conductor joined nerve and muscle, though less strongly. From these last results Galvani concluded that he had discovered "animal electricity," or, more exactly, the true nature of the ever-mysterious animal spirits. Electric fluid in the animal's nerve, led to the muscle by the conductor, caused it to contract. At this point Galvani went off into physiological and medical speculations of no great importance. Everyone knew that some animals (like the electric eel, *Torpedo*) could give violent electric shocks; Galvani was thus readily deceived into supposing that he had detected a more normal level of electrification common to all animals.

About a year after the first account of Galvani's experiments was published Alessandro Volta (1745–1827) of Pavia put forward a very different interpretation of them (1792). He could find no electricity in the frog's leg. For the convulsions to appear, a bimetallic contact was essential;* moreover, a circuit through the nerves alone would produce them. Volta asserted that the frog's leg was merely a delicate instrument for detecting electricity whose source was in the bimetallic junction. Soon he showed that the metals could be arranged in a definite order according to their polarity on contact, that any moist material would serve in place of animal tissues to produce the flow of current, and that the contact charge could be detected by the ordinary gold-leaf electroscope. This was a clincher for the identification of frictional and "animal" electricity. When Volta had discovered how to improve his electrical cells and link them into a battery, sparks and other effects made this identification very obvious.

Volta had not only revealed the first new source of energy

* Volta attributed Galvani's feeble contractions with a single conductor to impurities in the metal.

to be found in two thousand years,* he had demonstrated very dramatically that chemistry is an electrical science. In this, Volta, a physicist, displayed little interest, denying (mistakenly) that chemical action played a major role in the generation of electricity by his cells. Others seized energetically on this point, conceptually so fascinating, experimentally so fertile.

"Volta's pile" and his unwitting discovery of chemoelectricity intervened just at the moment when new light upon the nature of chemical combination was urgently required. For the facts themselves had recently crystallized into a new pattern that was clear, quantitative, and simple—yet inexplicable.

Although the writings of Robert Boyle (in the first place), Newton, and their followers had introduced into chemistry the principles of the mechanical philosophy, these principles had by no means won, in that science, the unassailable position they acquired in physics. Chemistry was not, and could not yet be, mathematical. No other logic served instead of mathematics to connect the assumed microphysical events—the dissolution and re-formation of particulate aggregations—with the experimental phenomena of chemistry. Older modes of thought, moreover, persisted either directly or by reflection in mechanical hypotheses. Despite Boyle's attack in *The Sceptical Chymist* (1661) the four Aristotelian elements were not utterly disgraced. The Flemish medical-chemist, Johann Baptista van Helmont (1580–1648) had revived Thales' idea that all substance is made of water; this was partially approved by both Boyle and Newton. Others saw all chemical reactions as typified by the acid–alkali antithesis. Most important of all, perhaps, was the idea that substances can be arranged as types, adding to the three alchemical principles (salty, sulphureous, and watery types), the earthy, the acid, alkaline, and metallic. Chemical change was supposed to occur between the different types in accord with vaguely defined patterns. Unlike the mechanical philosophers, most chemists of the seventeenth century were little concerned for theoretical ratiocination; after treating ideas

* Until about 1860 the chemical cell was the only useful source of electrical energy. Therefore most scientists doubted whether the application of electricity (in motors, for illumination, and so forth) could ever be economic, though battery electricity was economic for telegraphy, plating, and other special applications. Until near the end of the nineteenth century no one foresaw the immense utility of electricity in the convenient *distribution* of energy obtained from coal, oil, or falling water.

summarily, they went on straightforwardly to their main business, the systematic compilation of cookbook techniques. Since chemistry was practiced for the most part as an auxiliary to medicine, chemical textbooks were manuals on the preparation of medicaments. Philosophic chemists like van Helmont and Boyle wrote differently. But even they, and almost all other chemists, were affected by the strange delusions about the transformation of substances that were part and parcel of chemical mysticism, partly derived from alchemy, partly derived from Paracelsus, the founder of medical chemistry.

Most of the new compounds discovered in the seventeenth century, and there were many of them, were fruits of unsophisticated empiricism, such as were bound to turn up when a large variety of substances were heated, dissolved, distilled, crystallized, and so forth in more or less arbitrary mixtures. They were given bizarre names like "burning Spirit of Saturn" * (acetone, an organic compound), indicating perhaps something of their mode of preparation but rarely anything of their true composition. Many, like this, were obtained thanks to the now common use of the strong mineral acids as reagents. Many also, including sodium sulfate (the "miraculous salt" of Johann Rudolph Glauber) were speedily applied to medicine, usually with less effect than Glauber's salt. Boyle was remarkable among chemists because, although like others he profited from endless happy accidents, he followed a rough general plan—the investigation of the effect of the texture of things upon their chemical properties—and because he chiefly rejoiced in new observations that gave theoretical insight. His technical knowledge was superb because he had given serious attention to problems of analysis and of the distinction of one substance from another. Thus he had usually a shrewd and roughly correct idea of what happened to the "metallic part," the "acid part," the "earthy part," and so on, of his materials during the course of a reaction. So great was his skill and experience that he discovered very rapidly how to isolate phosphorus from the mere hint that the source was "somewhat to do with man's body" (urine, in fact). Yet for all this, and despite his far-sighted physicochemical outlook, Boyle was unable to give chemistry laws or system. No more was Newton.

After their day the science took a fresh turn, though it

* The metals were traditionally associated with the planets: Venus = copper, Mars = iron, Jupiter = tin, Saturn = lead, the Moon = silver, the Sun = gold. Of all these only Mercury still survives.

remained intensely empirical; indeed, empiricism was more strongly reflected than ever in its exponents' ideas. The metaphysical position of most eighteenth-century chemists before Lavoisier may be roughly defined by saying that they accepted the ultimately particulate nature of matter while imagining this structure to be codified either into Aristotelian elements or into the types already mentioned. They regarded chemistry as concerned only with these intermediate species, not with the associations and dissociations of the ultimate particles. Going further still, they began to discriminate between individuals within the types, recognizing (for example) that the "acid part" in saltpeter, say, was different from the "acid part" in copper vitriol, as also of course that the nonacid components of these two substances were also distinct. This implied, not only an obvious recognition of the derivation of these salts from the starting reagents, but that they were formed by a combination of the *parts* of the reagents that were also distinct in the first place. To some extent this point of view was already held in the seventeenth century, when the phenomenon of "double decomposition" was first attributed to chemical reaction, and Boyle clearly held the concept of characteristic parts that entered into a variety of combinations.* But the eighteenth-century chemists carried further their willingness to admit as the participating entities in chemical combination either substances considered to be simple (water, metals, sulfur, charcoal, acids, alkalies) or the parts of those thought to be compound, like niter, the vitriols, and salts. Thus, following a distinction made by Boyle, copper vitriol could be described as composed of a copper part and a specific acid part, iron vitriol being the same with iron substituted for copper. This was a step towards placing empirical knowledge concerning analysis and synthesis above all a priori theory concerning particles or elements.

Upon this development, which began to make a straightforward account of some chemical reactions possible, was superimposed a different, theoretical development which certainly also raised the level of intelligibility, for a time. To understand it, it must be recalled that in the early eighteenth century the complex of phenomena associated with the words

* Double decomposition may be expressed symbolically as the hypothesis that two compounds (AB) and (CD) sort out their components to form two new compounds (AD) and (B*C*). Glauber's account of the reaction used to prepare antimony trichloride may be thus rationalized: $Sb_2S_3 + 3HgCl_2 = 2SbCl_3 + 3HgS$.

fire–heat–flame–burning was not in the least well explained, and further that the study of substances was impeded by a complete ignorance of even the common gases. The idea that some substances could exist independently only as an "elastic fluid" was unheard of, and the "air" properly speaking was always regarded as chemically inert. The object of the phlogiston theory was to settle all these difficulties at one stroke. This theory originated in speculations of the mystic German, Johann Joachim Becher (1635–1682), alchemist and seeker after perpetual motion. Coherent form and the name "phlogiston" were conferred upon Becher's speculations by a German chemist, Georg Ernest Stahl (1660–1734), whose views were not widely accepted outside Germany until some years after his death. The influence of the phlogiston theory, greeted in France and England as the first great chemical synthesis, was of shorter duration than is often supposed; it was hardly fully established before its demolition began. Its initial repute was justified, for Stahl was a good practical chemist and his theory explained many reactions; it stood to Lavoisier's chemistry in much the same inverse relationship as Ptolemy's astronomy to the truer system of Copernicus.

In a sense phlogiston was "negative oxygen," necessarily so because Stahl had no notion of gases. When a combustible burned, or a metal was calcined, the substance gave off phlogiston (through the flame) to the atmosphere. Reduction of a metallic calx (oxide) necessarily required the restoration to it of phlogiston from a rich source, such as charcoal. This was the simplest and most credible part of the theory, which was also, however, extended usefully to many reactions other than combustion in which a chemical change could be attributed to a loss of phlogiston on one side and a gain on the other. But phlogiston never dominated the whole of chemistry. To many discussions it was quite irrelevant, since its natural fitness was for the explanation of exothermic processes.

Sometimes qualified as the "matter of fire" and always in close conceptual proximity to the "matter of heat" (Lavoisier's caloric, a transmogrified phlogiston having physical properties but lacking chemical ones), phlogiston as an impalpable, weightless fluid was no more absurd than other members of that group. As Bishop Watson said: "Show me a specimen of gravity or electricity and I will show you a specimen of phlogiston." The peculiar status of phlogiston created no problem until evidence was found to count against it. This evidence—the increase of weight in metals as they are calcined—had long ago been reported for lead, but it was not taken to be true of all metals until 1770. The moment was a critical one, for the progress of other chemical experiments was also threatening, not yet to cast doubt on the phlogiston theory, but to involve it in self-destructive complications.

Mayow had performed experiments over water; Boyle had collected "airs" (or elastic fluids) *in vacuo;* Stephen Hales (1679–1761) devised the pneumatic trough for the collection of "airs" from a number of substances. An experimenter much influenced by his notions of Newton's methods, Hales concluded that "airs" were fixed in solid bodies, from which they might be released by appropriate chemical processes. Twenty-five years later (1756) Joseph Black decisively identified such a fixed air as the essential substance that makes the difference between limestone and quicklime. Hales's work was well known on the Continent, Black's was not, otherwise the importance of Black's arguments based on *changes in weight* might have been appreciated earlier than it was by the French chemists. Black had neither isolated nor collected "fixed air." He had inferred its existence from reasoning and experiments precisely analogous to those from which the nonexistence of phlogiston in metals was to be inferred. Meanwhile, the investigation of gases was pushed further by Henry Cavendish, who isolated hydrogen and distinguished it from Black's "fixed air" (carbon dioxide); by the Swedish chemist Carl Wilhelm Scheele (1742–1786) who prepared both nitrogen and oxygen; and by Joseph Priestley (1733–1804) who independently discovered oxygen and a number of other gases. By about 1770, and still more in subsequent years, the phlogistic theory of these gases, some described as "phlogiston," others as "phlogisticated" or "dephlogisticated" air, was becoming wildly inconsistent. There were not enough qualifications of phlogiston to cover all the qualitatively different gases.

Although chemical researches in France at this time seem less dramatic than those elsewhere, they were notable for concentrating upon the crucial question of combustion and for laying the same emphasis on weights that Black's reasoning had shown. The French, too, seem to have been looking for explanations while the English—Priestley especially—looked for phenomena, to which they were content to attach any phlogistic rationalization that came to hand. Chemical logic was an outstanding consideration to Antoine Laurent Lavoisier (1743–1794), still a junior member of the Royal Academy of Science in 1770, but nevertheless inspired by discovering that combustibles increased in weight on burning, like metals, to embark on a complete study of the problem of combustion.*

The key to the increase of weight was, of course, the absorption of air during combustion or calcination. Adopting the Hales–Black concept of "fixation," Lavoisier showed in 1772 that a large quantity of a "fixed air" was evolved when lead oxide was reduced to lead with charcoal; later this was identified as carbon dioxide. Lavoisier's attempts to frame a comprehensive hypothesis for this fixation of air were, however, fruitless until, in 1774, he learned of Priestley's discovery of oxygen. The Englishman's qualitative observations upon this gas immediately suggested that *this* was preeminently the agent that supported combustion (and respiration); in fact, combining with calcined mercury it formed mercuric oxide, from which it could be recovered by strong heating, or, when the oxide was reduced with charcoal, it combined with the charcoal to form carbon dioxide.

The experiments performed by Lavoisier in the next few years fitted beautifully into the developing structure of his theory; step by step every fresh hypothesis predicted as the theory obtained depth and scope was confirmed by careful work in the laboratory. Not until 1778 did Lavoisier cease to entertain the phlogiston hypothesis, and he launched no attack upon it until 1783. By then he was ready for a thorough theoretical reform of chemistry, completed in his textbook *The Elements of Chemistry* (1789). Already a few French chemists had adopted Lavoisier's views, while within ten years only a very few diehards (Priestley and Cavendish among them) still clung to phlogiston. It was, like Newton's, a great triumph that could not have been won without the cooperation of many men, not least of the English empiricists who remained unconvinced of the truth of the new theory.

The new chemistry was distinguished from the old in four ways. It left to phlogiston only a physical significance as caloric. It reestablished the concept of element, though in a new form ("that which chemists cannot analyze"), and made the elements (plus caloric) the parties to chemical reactions: thus for example water consisted of nothing but hydrogen

* This discovery depended upon taking into account the gaseous products of combustion.

and oxygen, lead sulfide of nothing but lead and sulfur.* It conceived of the gases as elements or compounds, like the liquids and solids. And it established firmly the notion of equivalence in chemical change—the weight of the products must be equal to the weight of the reagents. Thus the open end of all previous chemistry, created by ignorance of the role of gaseous substances, which phlogiston had failed to close, was finally sealed off. Naturally the new also differed from the old in the description of every chemical phenomenon, often in the sense that a translation differs from an original, for Lavoisier revolutionized the concepts and intellectual analysis of chemistry, not its empirical data. His function as an innovator was (once more) like that of Corpernicus or Harvey rather than like that of Newton or Darwin. Perhaps a fifth difference should be added: the new chemistry was not a theory of matter. In some measure at least all earlier speculations had sought to postulate an a priori theory of what matter is and what its fundamental properties are, then to show that chemical facts were accounted for by this theory. That had been Boyle's ambition, but it was not Lavoisier's, not that of a logically gifted empiricist who revealed little interest in the metaphysics of nature. Lavoisier did not explain how the elements combine, nor why only some combinations are possible. He had no views on the ultimate reasons for the qualitative differences between the elements. He offered no explanation of the fairly large body of empirical data concerning the proper proportions of reagents (by weight or volume) to be employed in a given process. Perhaps he would have cast light upon these questions, but his scientific work was brought to a close by the demands of the French Revolution, which, within a few years, cut off his head.

Lavoisier's death left, anomalously, a new chemistry without a physical substratum. If less occult, the position was intellectually scarcely less puzzling than it had been when chemistry was a mystic science. The facts of chemical combination were well ordered at last, but the breach between the chemists' language and view of nature and those of the physicists was now complete. Temporarily at least, under

* The modern chemical nomenclature (here often used anachronistically for convenience, along with some of the older names) was introduced by Lavoisier and his colleagues in 1787. Its principal objects were to introduce new, clear names for the elements where required (oxygen, hydrogen) and to indicate in the names of compounds their elementary composition, according to a systematic plan.

Lavoisier's guidance, the chemists had renounced the mathematical, mechanical metaphysic of nature that had achieved so much, returning to a theory of real qualities (for such the distinctions between the elements now were) that Galileo had rejected almost two hundred years before.

Hence the importance of Volta's discovery of chemo-electricity: it disclosed the possibility that, after a short divorce, the intellectual worlds of physics and chemistry might be reunited. For this must follow if, as now seemed to be the case, the bonds that held the elements together in chemical combination were of an electrical nature. In 1800, soon after Volta's famous letter describing his "pile" reached London, Nicholson and Carlisle dissociated water by running a current through it. Before long Humphry Davy (1778–1829), also by electrolysis, had isolated two new elements, immensely reactive, sodium and potassium. It could hardly be doubted that chemists were, after all, unwitting and reluctant experimental physicists.

Chapter 11

THE BEGINNINGS OF BIOLOGY

The seventeenth century gave physical science a new framework of ideas, that of mechanism; it demonstrated in Newton's brilliant accomplishment the power of a new logic, that of mathematics; and finally it established the principle that physical theories should be capable of verification by observation and experiment. But the seventeenth century did not witness a similar total change in the character of biological science—indeed, the notion that all aspects of the study of living things constitute a unified field of knowledge failed to emerge. Nor was there that close relation between abstract thought and scientific investigation that held good for the physical sciences. Men like Bacon, Galileo, Descartes, Boyle, and Newton were, at least in part, philosophers in the most general sense; this was not true of the naturalists, microscopists, and anatomists. The activities of the latter group did not force them to postulate general truths about the nature of things analogous to those of the physicists, and, in so far as they needed to appeal to such truths, they were content to accept without criticism age-old dogmas concerning the changelessness of living species and the eternal ladder of nature. By the early eighteenth century all non-rational elements had been banished from the physicists' discussions, while the problem of creation (or *Genesis*) inevitably inhibited the wider speculations of biologists for more than another hundred years.

The fact that biology was devoid of even one general scientific hypothesis until 1838 does not mean that it was sterile; it did insure that the advances were of limited scope, and unrelated to one another. The history of modern biology before the nineteenth century lacks the internal consistency of the histories of physics and chemistry. In particular, original speculation with regard to living things—of which there was plenty—tended to be highly personal and to depart wildly from the important developments actually taking place at the descriptive level. For example, the revelations of the microscope were no less exciting and stimulating than those of the telescope a couple of generations earlier, but their incorporation into a coherent, demonstrable body of scientific theory took far longer. Of course, the phenomena were much more complex.

Although many, even in the seventeenth century, entertained the ambition to extend to living things the same metaphysics and the same methods that were proving so promising elsewhere, they met with but partial success. Physiology offered the best ground for this ambition. It had long been clear that in some obvious ways (such as the action of muscles to move bones) the animal body can be treated as a machine; and the distinction between chemical processes (occurring in vitro) and physiological ones (occurring in the organism) had been confused, or considered as providing useful analogies, even by Galen—not to say Paracelsus. In limited respects a reinforcement of the physicist's and chemist's approach to physiology did enjoy some triumphs.

The simplest way to consider a living creature mechanically is at once to deny, in principle, that it possesses any non-mechanical attributes, such as sensation or the power of choice. This drastic step was taken by Descartes, who was careful, however, to distinguish man from animals by his possession of a rational soul. While ancient doctrine taught this too, it had maintained that other characteristics of life belonged, in decreasing measure, to humbler creatures than man, forming in their successive order a "chain of being." Similarly, this chain stretched upwards, above men (the humblest reasoning creatures) into the intellectual hierarchy of angels. Thus (as Lady Chudleigh wrote in 1710):

> We have reason to believe that as we see an innumerable company of beings below us, and each species to be less perfect in its kind until they end in a point, an indivisible solid; so there are almost an infinite number of Beings

> above us, who as much exceed us, as we do the minutest insect or the smallest plant. . . .[1]

Life was held to be a continuous spectrum, with each species assigned at the Creation to its proper and unalterable place and with no niche left unfilled by the Creator. By contrast, Descartes taught that all plants and animals (including the body of man) were mere machines, just as the universe was a great machine, performing some functions like digestion automatically and responding to events in their environment in predetermined, mechanical ways. A man, having a soul, could *feel* pain; the cries and struggles of an agonized animal were no more than reflex actions.

As might be expected, Descartes' chief concern in mechanistic physiology was the nervous system; he contributed little or nothing to speculation about those processes which were already quite commonly regarded as chemical in nature. In his *Treatise on Man* (posthumously published in 1662) he set out for the first time the notion that a signal, carried to the brain by a sensory nerve from the periphery of the body, could automatically trigger off certain responses. Through purely mechanical linkages in the brain, appropriate instructions would be dispatched along the motor nerves to muscles so as to cause them (for example) to move a limb that experienced pain, to utter a cry or to weep, and to turn the eyes towards the source of the pain. In some cases (as in the automatic regulation of the thoracic muscles responsible for breathing), this exchange between sensory and motor nerves might take place outside the brain in some subordinate center. Assuming the machine–organism to be sufficiently complex, every one of its capabilities except rational thought—which Descartes took to be indeterminable—could be treated as built into it from the first. Within a completely mechanistic physiology, terms like *spirit* or *soul,* implying the existence of some unknowable, immaterial agency directing the organism, were meaningless and unnecessary. Believing that man alone possessed a pineal gland (at the base of the brain), Descartes assigned to it the seat of the human soul, in a position where it could consciously control, by will, the operation of the nervous system.

Despite all the crudity of Descartes' model, if we may reexpress his thesis as stating (1) that all biological theories based upon souls, life-forces, or other unknowable entities are scientifically worthless; and (2) that scientific biology consists of the analysis of living phenomena into physical and chemical processes, then Descartes enunciated the principles of nineteenth- and twentieth-century experimental biology. Many biologists of recent times have indeed held that ultimately life cannot be reduced to physics and chemistry and they have pointed out (properly) that the degree of organization of living things is such that their behavior becomes qualitatively distinct from that of their component physical particles. Hence living things are subject to their own biological laws, such as that of evolution. Nevertheless, biologists have generally recognized that the first step towards investigation involves making the assumption that Descartes first introduced—that living nature may be understood by analytic methods since it contains no ineffable mystery.

Although the effective use of mechanical arguments in physiology had been revived by Harvey,* an anatomist and a medical man—from whom Descartes gladly borrowed for his own account of the cardiovascular system—and although mechanical explanations of some functions or attributes of animals became standard in biological literature, the degradation of all nonhuman life to automatism met with a cold reception. In this as in all other respects Descartes was not without followers, but working biologists were little influenced by their speculations. For it was one thing to liken an unconscious function such as digestion to a chemical reaction or to fermentation—this was understandable—and quite another to reduce all the seemingly innate responsiveness of an animal to the operation of mechanical linkages, especially as this responsiveness seemed to reside even in the animal's organs and tissues. True, some seventeenth-century investigators like Francis Glisson (1597–1677) postulated in organic tissues an inherent "irritability" that was the origin of the organ's responsiveness, which Albrecht von Haller (1708–1777) was to make the basis of animal physiology. But irritability was a vital property that defied the mechanists, even though Julien de la Mettrie (1709–1751) made use of it in order to extend the mechanistic principle from the physiological level, where Descartes had left it, to the psychological. La Mettrie's *L'Homme machine* (1748) denied the existence even of the human soul. Again, la Mettrie's arguments (though they enraged theologians and all right-thinking men) were of no relevance or interest to working biologists, who continued to draw their metaphysics from

* Galen had on occasion used mechanical arguments not dissimilar to Harvey's—e.g., in determining the flow of urine from the kidneys to the bladder.

conventional sources and very often to assign distinct living principles to the smallest particles of organic matter. It was respectable enough to compare the universe to a clock, yet naturalists, philosophers, and theologians joined in protest against the suggestion that any *living* portion of this universe was mechanical. Descartes had stretched the seventeenth-century metaphysic of science beyond its breaking point, rendering biological mechanism so stark and repugnant that it was firmly rejected, as a general principle, by all but a very few.

When directed to particular problems such as digestion and respiration, a more restricted physiological mechanism seemed perfectly reasonable. Among the members of the Royal Society in London suggestions concerning the chemical mechanism of respiration first put forward by Boyle and Hooke were developed through physiological experiments. Further analysis of the cardiovascular system led the physician Richard Lower (1631–1691) to connect the change from venous to arterial blood occurring in the lungs with the passage into the blood of a "nitrous" spirit derived from the air. Experiments with the air pump and on artificial respiration all seemed to prove that the physiological significance of air in breathing was akin to, and probably identical with, its chemical significance in burning; in John Mayow's writings there is a fairly clear suggestion that the function of the arterial circulation is to transport "nitro-aerial" particles from the lungs to the tissues of the body. With Mayow, even more than with his predecessors, the problem of breathing is reduced to purely physical terms in an elaborate form of mechanical hypothesis. Correspondingly, when combustion was first accounted for in a wholly chemical context by Lavoisier, it became relatively easy to understand the oxygen–carbon dioxide exchange effected by the blood. (Significantly, the former gas was first identified by Priestley as "eminently respirable air" while Black had earlier discovered his "fixed air" in the exhaled breath.) The investigation was taken to a further stage when Lavoisier and Laplace demonstrated experimentally that the heat produced by an animal from a quantity of food is equal to that obtained by burning the same amount of food.

Important as these new developments were they left untouched the central problem of physiological mechanism—the understanding of precisely how organisms effect chemical and physical changes. It by no means follows (for instance) that an oxidation occurs in the body tissues exactly as it does in the chemical laboratory. Since the delicate methods of biochemistry were not available in the seventeenth and eighteenth centuries the analogy between living and nonliving processes could only be grossly stated in terms of input and output. How deceptive such a crude approach could be is obvious from the repeated experiments which seemed to show that plants required nothing but water for their growth, contrasting with others that, by destructive distillation and analysis of their ashes, showed plants to be composed of "air," "oils," "salts," and "earth," as well as water. The direct approach to the study of vital function had to follow, at this stage, other means than the chemists'.

An experimental procedure that was still valid, and long remained so, was the classical one of vivisection. By simple surgical procedures Hooke and others proved the independence of the respiratory function of the thoracic movements, while Regnier de Graaf (1641–1673) attempted by extirpating the pancreas from dogs to demonstrate the uses of that organ. But it was not until after seventeenth-century attempts to investigate digestion by such means had proved ineffectual that studies of the actual processes in the alimentary canal were begun by R. A. F. de Réaumur (1683–1757)—no less considerable as a biologist than he was as a physicist. Réaumur enclosed meat in perforated tubes which he caused a kite to swallow, then regurgitate; using sponges he collected samples of the gastric juice. A similar technique was used later by Lazzaro Spallanzani (1729–1799) who succeeded in demonstrating partial in vitro action of the juice. Such experiments appeared to confute van Helmont's comparison of digestion to fermentation, but the actual chemistry of the body fluids defied analysis as yet. Greater awareness of the body's chemical complexity, taught by such investigations as these, tended through a century and more to cast doubt upon the seventeenth-century's assurance that physiological functions could be accounted for in chemical terms, as well as to strengthen the conviction of the vitalists that living organisms could both break down and synthesize substances in ways that the chemist could not duplicate. The laboratory preparation of an organic substance, urea, by Wöhler in 1828 is often regarded as the first breach in this position; however, Wöhler's report did not carry this significance to contemporaries, and in any case he employed a simpler organic material (ammonium cyanate) in effecting his preparation.

Much physiological experiment was directed towards medical practice. An obvious example is the series of experiments made by Christopher Wren and others on the effect of drugs when injected into the blood stream of animals. (The successful medical use of intravenous injection was delayed until the nineteenth century.) From these experiments developed the extravagant ambition to try the effect of transferring the blood itself from one creature to another. At last, in London in 1667, the blood of a healthy sheep was (supposedly) substituted for that of a "poor and debauched man, cracked a little in his head." When the subject of a similar experiment died in Paris soon afterwards the whole question was abandoned for almost two hundred years.

Although physicians had long used variations of body heat and pulse rate in diagnosis, neither thermometers nor accurate clocks were adopted into medical practice during the seventeenth century when these instruments first became available. Apart from the investigations associated with the air pump, experimentation in physics had little effect on physiology until Stephen Hales (1679–1761) examined the pressure of fluids in plants and animals. The former experiments (*Vegetable Staticks,* 1727) were the more important, for Hales revealed the relatively enormous pressure of rising sap—still not fully accounted for—and measured the rate of transpiration through the leaves. These, like Hales's other experiments on the yield of "airs" from a variety of organic substances, were "input and output" experiments, not wholly dissimilar in principle from those which Sanctorio had made a century before in his weighing-chair upon the intake and excretion of food from the body.

During the whole period of early experiment in physiology anatomists had continued their more traditional examination of structure, which inevitably (as with the discovery of the lacteals and the thoracic duct in the second quarter of the seventeenth century) had profound effects on the study of function. Although the microscope was invented almost concurrently with the telescope (in 1609) anatomical research virtually ignored it until the early nineteenth century. Effectively, the microscope did not become serviceable in biology until about 1660 and even then it was little employed for the observation of fine-structure. The first major book on microscopy (which derives most of its importance from its splendid plates) was Hooke's *Micrographia* (1665); while making some structural observations (of the sting of the nettle, for example) and observing cellular structure in cork, Hooke had no inkling of the value of the new instrument for physiological purposes. Few anatomists or naturalists were more prescient. The conspicuous exception was Marcello Malpighi, who not only discovered the capillary circulation but went on to employ the microscope for histological examination of the tongue, liver, and other organs, as well as for following the development of the chick embryo. Further (with the Englishman Nehemiah Grew), Malpighi opened up the whole subject of plant anatomy, the necessary key to plant function. Through their combined efforts this aspect of botany was raised, in a few years, from nothing to the point where fine-structure in plants was better described than that of animals. Thus Malpighi was to some extent justified in his hope that plant anatomy, being simpler, would throw a clearer light on the microscopic anatomy of animals.

Despite such observations as Hooke's of cork cells, Grew's of "bladders" in plants, and Leeuwenhoek's of red blood corpuscles, the microscope at this point in its development was too imperfect an instrument to permit the study of cellular structure, and the theory that cells form the universal elements of living architecture was delayed for another hundred and fifty years. Cytology was a product of the optical improvements in the microscope that were made during the first quarter of the nineteenth century. Seventeenth-century optics was adequate, however, for the study of insects, their anatomy, life cycles and forms of reproduction, a field of natural history greatly neglected in the past. Each of the three great microscopists—Malpighi, Jan Swammerdam (1637–1680), and Antoni van Leeuwenhoek (1632–1723)—added fresh knowledge, making it clear that insects (not to mention such creatures as frogs and lobsters) were organisms no less intricate in their functioning than animals, and furthermore were "designed" after plans very different from those of animals. Entomology suggested significant, if not always correct, ideas that passed into general biology. In the first place, the notion of spontaneous generation was (at this scale) decisively confuted, since all these species had particular and elaborate reproductive systems. The same point was made upon experimental evidence by Francesco Redi (1626–1698), a distinguished member of the Accademia del Cimento: meat protected from flies did not produce maggots. A doctrine handed down from Aristotle, which the naturalists of the Renaissance had not hesitated to endorse, was thus reduced to the level of a mere superstition, although, in different form, it was to be revived to trouble Spallanzani in the eighteenth century and Pasteur in the nineteenth. Less fortunately, entomology also confirmed the theory of "preformation." Many insects and amphibia underwent a metamorphosis through which, apparently, an invisible, preexistent form emerged to the light of day. Accordingly Swammerdam, for instance, insisted that the butterfly preexisted in the caterpillar. Aristotle had taught that development involves creation of new parts, that is, the occurrence of differentiation in undifferentiated matter. In this he had been followed by Harvey. The microscope, however, confirmed the belief that the embryo was preformed in the egg, complete and perfect though of submicroscopic dimensions, so that no creation of new structures occurred during development but only enlargement of existing ones. Indeed, there was a natural correlation between mechanism and the preformation theory, for it seemed that only a constant miracle could explain the organization of a chick out of nonchick material—epigenesis made life an unfathomable mystery. The preformationist could at least teach the consistent, if numerically fantastic, theory that the creation of *all* living beings had occurred once, and once only, at the beginning of the world, and that nothing but a mechanical unfolding of these invisible embryos, passed from one generation to the next, ever happened again through the whole history of the world. Incidentally, this theory of generation fully agreed with the naturalists' dogma that living species are unchanging and unchangeable.

Preformation remained the dominant concept in embryology till about the end of the eighteenth century. Usually the embryo was thought to reside in an egg, for there was no great novelty in William Harvey's motto that "all living things come from eggs." As nothing like a mammalian egg had ever been observed until De Graaf reported it in 1672,* the mammalian ovary being considered a *testis* or gland, this notion was somewhat premature; more logically, naturalists had long made a prime distinction between viviparous and oviparous animals. The simple consistency allowed by De Graaf's observation, making all creatures truly egg-descended, was upset in a few years by Leeuwenhoek's account of the spermatozoa in semen; he at once propounded the theory that these lively animalculae contained the embryo, and that the egg (into which he assumed the sperm penetrated to effect fertilization) merely provided shelter and food for it. Neither the "ovist" nor the "animalculist" pre-

* Properly, the follicle still bearing his name. The true ovum was first seen by von Baer in 1827.

formationists paid the least attention to the obvious facts of heredity or to the experience of breeders, which Darwin first brought into biological discussions. This controversy again continued well into the nineteenth century, when the physiology of reproduction at last assumed a coherent form on a firm basis of experimental and microscopic evidence. Indeed, until 1820 the notion that the sperm had any essential part in the process was very widely doubted, even by Spallanzani, who actually provided some experimental proof that they did. His skepticism was no more remarkable than that of Malpighi, who firmly asserted the preformation of the embryo, despite his own fine drawings clearly indicating development!

These microscopic researches in embryology, histology, and comparative anatomy are clearly connected with traditional problems upon which (until the beginning of the seventeenth century) Aristotle had remained almost the only, and the undoubted, authority. To these the microscope added new dimensions of depth and time. Both Malpighi and Swammerdam were medical men (as were many other early microscopists), and their work, for all its originality, bears the mark of their medical training in anatomy and physiology. The most astonishingly unexpected realm of vital activity was, however, opened up by Leeuwenhoek, an unlearned man with no language but his native Dutch, who communicated all his observations to the Royal Society in London (as did Malpighi, for the most part). Attracted by chance to the study of the minute, Leeuwenhoek acquired an extraordinary talent for grinding tiny single lenses, which gave a magnification as high as 400 times. With these he examined, as his fancy took him, an almost endless series of objects. This wide-ranging curiosity led him in the summer of 1674 to "view" with a microscope some greenish water from a pond near Leiden:

> I found moving in it several earthy particles, and some green spiral streaks like the "worms" of copper or tin used by distillers to cool their distilled liquors; and the whole compass of each of these streaks was about the thickness of a man's hair . . . among all which there crawled abundance of little animals, some of which were roundish; others, somewhat bigger, were of an oval figure. On these latter I saw two "legs" near the head and two little "fins"

on the other end of their body. Others were somewhat larger than an oval, and these were very slow in motion and few in number. These little animals had various colors, some being whitish, others transparent; others had green and very shiny little "scales," others again were green in the middle and white in the front and rear; others were grayish. And the motion of most of them in the water was so swift and so various . . . that I confess I could not but wonder at it. I judge that some of these little creatures were more than a thousand times smaller than the smallest ones I have previously observed in cheese, wheat-flour, mold and the like.*[2]

Thus Leeuwenhoek announced the discovery of single-celled organisms, protozoa—not knowing, of course, that they constituted a new order of animals and always supposing that they possessed limbs and organs like the metazoa. During a half-century of microscopy Leeuwenhoek described many such organisms, which he found almost everywhere, noting that some of them could withstand long periods of dehydration. He also observed a few species of bacteria, both free-living forms and parasitic species such as those of the digestive tract. Unfortunately, there was no framework of biological ideas or techniques within which Leeuwenhoek's observations could be incorporated; moreover, other scientists lacked instruments capable of repeating and extending them. True, the microscope had disclosed a new, unsuspected level of existence (which was sometimes hypothetically linked with the occurrence of disease) but no one knew how to study it, or to draw from it any general conclusions about the living state. The "little animals" were generally classed with far larger, and very different, forms, such as the already well-known "vinegar eels" (nematode worms). The best that could be done was to study them in a "natural history" way, and after Leeuwenhoek's death interest focused, from this point of view, on more accessible creatures (like the polyps) though the description of some protozoa (such as *Paramecium*) was certainly much improved. The most interesting microscopy of the mid-eighteenth century was that of Abraham Trembley (1700–1784), working with the fresh-water polyp *Hydra,* in which he discovered regeneration and asexual reproduction. But closely linked with the cell concept

* The organisms described here have been identified as algae, rotifers, ciliates, and flagellates.

as it must be, scientific protozoology was to be of nineteenth-century foundation.

Since 1800 the biologist has been, primarily, a microscopist. The microscope was the constant tool of all the great nineteenth-century figures—not only of the cytologists, but of Darwin, Huxley, Pasteur, and the rest. Between 1650 and 1800 this was far from being true, since the most active naturalists and anatomists made little or no use of this instrument. In many fields—taxonomy, comparative anatomy, paleontology, and so on—gross observation was equally important, and as many new ideas resulted from that as from microscopy. Thus the Dane Nils Stensen (1638–1686) and others were drawn to assert the organic origin of fossils from the clear resemblance of their *form* to that of contemporary teeth and shells, while other naturalists no less cogently argued from their *material* that they were mere chance stones. Evidence, concepts, and time scale were all inadequate to substantiate the former view—for no one would yet admit that living species had evolved by modification from the fossil ones—so that the seventeenth century again failed, in biology, to cast more than a pale doubt upon traditional Biblical ideas of creation. If fossils were taken to be organic, they could be explained as vestiges of an antediluvian creation; if inorganic they could be regarded as mere "sports of nature." Either way the Biblical chronology could be preserved.

The naturalists received little help from the geologists towards liberating themselves from religious mythology. Although, following the example of Descartes, most writers on the history of the Earth were now willing to admit that it had passed through many vicissitudes (Steno had devised a "model" for the distortion of rock strata laid down in successive layers, and speculated that the rock matrix in which fossils are found was originally mud or clay) they compressed all this change into the 6,000 years of Biblical chronicle. Even the heterodox Buffon in the mid-eighteenth century was unwilling to multiply this period by more than about ten times. Within such a brief stretch of time biological change—if it had occurred at all—must necessarily have been catastrophically violent, indeed miraculous; it was easier and more consonant with the principle of uniformity to suppose that such change had never occurred at all.

The whole period from Harvey to Linnaeus was one in which the uniqueness of life was strongly emphasized. The most obvious case of this was the downfall of the concept of spontaneous generation; it now became quite inconceivable that nonliving could ever generate living matter. The essence of life became its transmission from parent to offspring, each generation being a replica of the previous one. The mechanistic philosophy, even in its application to functional physiology, did not in the least challenge this concept, while microscopy, natural history, and embryology combined to endorse it. Any variation between members of the same species was attributed exclusively to differences in environment. Specific identity arising from characteristics created at the beginning of the world, perhaps carried through a Noah's Ark pair, could never be modified by subsequent events. The mechanical philosopher found no difficulty in contemplating living things as unchanging, created cogs in the universal clockwork; since the Newtonian universe was perpetually the same, there was no reason why its inhabitants should not be so too. To naturalists, little troubled as yet by the fossil record, no other explanation of obvious ecological facts was required.

The old Aristotelian philosophy of matter and generation, a philosophy of "becoming," accounted inadequately for the tight, inviolable lines of specific descent which neither time nor miscegenation (apparently) could weaken. The "hard" pre-Darwinian concept of species solved the problem by defining as a living species that which is incapable of hybridization or any other cause of mutation. Each species could reproduce only its own kind, and Providence insured that none disappeared. To a large extent, however, theology only reinforced common sense, for the variation of species is difficult to observe except under peculiar experimental conditions. (Buffon was the first to point out that domesticated species of animals and plants do differ markedly from their wild relatives, but this could again be attributed to the effect of the unnatural environment provided by cultivation.) Faced with the facts that hens breed hens and never ducks, although every hen is not identical with every other hen, naturalists preferred quite reasonably to adopt the type of theory that cemented the genetic traits of being a hen into a single unbreakable whole; to have done less would have seemed to admit that a hen might one day lay a duck's egg. In denying that this could ever come about, the eighteenth century was anti-Lamarckian before Lamarck. Trembley, for instance, proved that monsters could be made by grafting hydra as one would graft apple trees, but such surgery did not modify reproduction of the species.

Yet with all this there was still doubt. However firmly created, however mechanistic, the universe, no one could be quite sure that there was no "plastic" power in nature that might produce future forms different from those of the past. This assumption of plasticity was, of course, to provide the foundation for nineteenth-century ideas of evolution. Competent botanists like Ray and Linnaeus knew that *some* hybrids are fertile, giving rise to varieties that (were their origins unknown) the taxonomist would class as true species. Hence Linnaeus, at the end of his life, admitted that numerous related species might have fanned out by variation from an original pair of organisms. In the latter part of the eighteenth century, reaction against the "mechanical," or Linnaean, sexual system of taxonomy implied no direct withdrawal from the concept of the fixity of species, but it certainly rendered it less "hard." A species was seen less as a creation of God than of the taxonomist. Increasing knowledge of the fauna and flora of the Americas, the Pacific, and Australasia created some awareness of the naivety of Garden of Eden mythology. Had God created peculiar creatures like the kangaroo in only one region of the globe, and yet separately created identical species more than once in far-removed regions? Was there no meaning in the similarity of structure constantly elucidated by the comparative anatomists, in the mammalian anatomy of the dolphin, or in the human characteristics of the chimpanzee, clearly brought out by Edward Tyson?

The answers to these and many other questions demanded that biology desert its three guide lines: Scripture, teleology, and the argument from design. Effectively, physiological studies of function had already done so by the end of the seventeenth century. But the character of natural knowledge as a whole was determined by natural history, where by the end of the eighteenth century even the staunchest advocates of the fixity of species had renounced the simplest interpretation of divine creation—without, however, also renouncing divine intervention. Rather they postulated that God had repeatedly made over his handiwork, like a bungling craftsman. Curiously, biologists did not understand that this kind of hypothesis had already been disposed of by Leibniz:

> According to my opinion, the same force and vigour remains always in the world, and only passes from one part of matter to another, agreeably to the laws of nature, and the beautiful preestablished order. And I hold, that when God works miracles, he does not do it in order to supply the wants of nature, but those of grace. Whoever

> thinks otherwise, must needs have a very mean notion of the wisdom and power of God.[3]

The physical scientists had long ago learned that the universe cannot be both law-governed and God-governed. The biologists had yet to learn the force of this exclusion. When once they had learned it and given full recognition to the increasing evidence that fossil species differed from those now extant, they would be compelled to admit organic evolution as a natural law and part of the "beautifully preestablished order." Catastrophic hypotheses insulted the divine providence no less than the human intelligence. If species had changed, then a foreseeing Creator must have designed their evolution—unless that too were to be mechanistically explained.

V

The Establishment of Science in the West

V

The Establishment of Science in the West

Chapter 12

SCIENCE AND SOCIETY

The tradition that scientists are abstracted, impractical men dies hard, and it is to be hoped that it will prove immortal. There is no reason to believe that men who have devoted part of their lives to physics or zoology make worse administrators and politicians than those who have been educated in classics or law, or nothing, and some grounds for thinking that they would do better; but they cease to be scientists. It is foolish to speak of the future, but one can say confidently of the past that so far every advance in science has been an intellectual advance, an increase of understanding, and that these advances have been made by men without thought for the usefulness or otherwise (in practical terms) of their activities. Even of the fundamental discoveries that made modern medicine possible this is true. It may be the case that the future development of science will depend less on sheer disinterested curiosity, and more on organized researches designed to shorten life by war or prolong it by medicine; of the past one can only say (with Francis Bacon) that experiments "of light" have always been more highly valued than experiments "of profit." Throughout antiquity and the Middle Ages, indeed, there was hardly any other expectation: the scientist was a philosopher whose business was not man, nor society, nor the search for abstract truth, but the study of nature for the sake of understanding it. Very little resulted from science to belie this expectation, for if it yielded greater understanding it brought little more power over events. The farmer, the craftsman, the soldier, and the sailor, pursued their own paths, expecting nothing from science and receiving little.

It is true that even in antiquity mathematics was applied to surveying, astronomy to navigation, and mechanics to the art of war, in the same way that the arts of poetry or history have been borrowed (some would say prostituted) for political or military purposes. But real mathematicians and real scientists were not misled into thinking that these applications were the objects of their intellectual inquiry into nature. Similarly, throughout the Middle Ages when the philosopher, with his strongly logical approach to scientific study, was perhaps more remote from the craftsman than at any other time, the very important technological progress of this period was the product of empirical invention, not science. So much was this the case that, in early modern times, Francis Bacon could contrast the solid inventiveness of the crafts with the sterility of philosophy (as he saw it). The success of empiricism in technology was grounds for believing that empiricism would work well in science. Therefore, Bacon said, science should approximate more closely to craftsmanship, and philosophers should not despise the knowledge that miners, fishermen, and artisans possess.

In return for such knowledge, science should impart direction and theory to technology. Bacon insisted again and again on the social relevance of scientific research: the very first aphorism of his *Novum Organum* runs:

> Man, as the servant and interpreter of Nature, does and understands things to the extent that he has studied the order of Nature both intellectually and practically; he neither knows more, nor can accomplish more than that.[1]

Intellectual understanding and practical accomplishment are for Bacon inseparable and interdependent. Scientific research supplies a reservoir which is potentially full of benefits to man, from which he can derive medical relief, liberation from harsh labor, and independence of nature's vagaries. Bacon's point of view seems too well known in its modernity to require amplification. But it should not be misunderstood. As Bacon emphasized the reform of logic as a desideratum even greater than an artisanlike empiricism, so in his eyes scientific knowledge came first, and utility flowed from it. He would have abhorred the doctrine that research should address itself to practical problems only, foreseeing that this policy ends in failure. For him, as for many later scientists, usefulness was not the object of science, but rather the justification for the encouragement and facilities that society should give to the scientist, so that he may add to knowledge. Bacon, it is clear from the *New Atlantis,* did not envisage a technocratic community but a scientocratic one, science in its best sense embracing learning and wisdom beyond the technical mastery of physics or zoology. Such a community has yet to be built.

Bacon's message was never forgotten and has been frequently restated, by novelists like H. G. Wells besides philosophers and sociologists. From the seventeenth to the nineteenth century it assured men with sensitive religious and social consciences that science was more than an idle intellectual pastime, for its beneficent results gave science a moral purpose. If, since 1945, such complacency seems naive, it should be remembered that war as a social cataclysm is a very recent phenomenon—no older than scientific warfare. Even hindsight can scarcely doubt that until 1900 science did far more social good than evil, or that until 1945 disease, starvation, and poverty (which science has alleviated) were greater dangers to civilization than war (which science has intensified). Before 1945 scientists, like almost everyone else, accepted war as a virtually inevitable tragedy in society. Like others, they succumbed at times to the glib fallacy that when weapons become more destructive war becomes "unthinkable"—or at least briefer—and comforted themselves with the pathetic truth that medical science has eased the sufferings of the wounded. In short, scientists, like other people, lacked wisdom and foresight and sometimes they shrugged off responsibility. *They* would provide the knowledge upon which the good society could be founded; let it be for society to decide how this knowledge should be used. Few have had Nobel's acuity of conscience.

Three centuries of advocacy of science have, at any rate, followed Baconian lines. Science is good because it is intellectually appropriate, because it promises advantages to mankind, and because it is a form of worship of God's majesty in his creation. Only the first of these sentiments would have been understood by the Greeks, but it is surely no accident that, until recently, science and religion have gone hand in hand. The wider, more speculative reaches of the European and American scientific tradition are (despite Laplace) thoroughly infused with the notions of design, order, and purpose. The philosophy of science—in so far as it is not logic—was no more exempt from them. Even in the nineteenth century, with the rare exception of a Thomas Henry Huxley, it was the humanists, not the scientists themselves, who brought the weight of scientific argument to bear against theology. Charles Darwin dealt Christianity perhaps the greatest blow it has ever suffered (though not the greatest it has ever received), but he never wrote a word hostile to religion. Quite the reverse; the scientific agnostics of the nineteenth century were apologists, not crusaders. They wished to render scientific thinking more acceptable than it already was, not less so. Perhaps social conformity is a necessity for science, since it enables it to survive. The group of men who met in Oxford and London under Cromwell became the Royal Society under Charles II. In France science passed from monarchy to republic and through Napoleon's empire back to monarchy again—though not without the loss of some heads. Today science flourishes (as every technical communication admits) equally well in Soviet Russia, or the United States, or Japan. Societies of utterly alien character have at least their dependence on science in common, perhaps belatedly recognized, and science serves each of them as though indifferent to creed or ideology.

Society's first recognition of the Baconian claims may be found in the foundation of the Royal Society (1662) and of the French Royal Academy of Science (1666). These events mark the first interest of the nation–state in knowledge as an instrument of policy. Probably Charles II expected little of what was, for him, an indulgent but inexpensive gesture, and he was happy to laugh at his philosophers for doing nothing but weigh the air. Louis XIV was the more deceived, since his academy neither strengthened his forces in war nor rendered his industries more lucrative in peace—at least not quickly. Before long there was a countermovement, in which the litterateurs satirized science for seeking to know only what wasn't worth knowing. That long, ponderous, solemn, yet depraved and corrupt age, the mid-eighteenth century, cast science into its own mold, as each age does; its epitome was Linnaeus. Science was rendered respectable, and taught (suitably simplified) to young ladies. Sermons and moral verses were adorned with allusions to Newton and nature. Dr. Johnson thought chemical experiments were a suitable diversion from serious reflection. Only among the dissident, witty French *philosophes* was the daring, unorthodox speculation of the seventeenth century continued. Nevertheless, in the industrial regions of Europe a change was working, and the alliance of science and technology was taking shape. Though the steam engine may be traced somewhat deviously to the seventeenth-century concept of atmospheric pressure, the beginnings of modern technology in the so-called Industrial Revolution of the eighteenth and early nineteenth century owed virtually nothing to science, and everything to the fruition of the tradition of craft invention. Such features of the Industrial Revolution as the factory system and the division of labor; the mechanization of the textile industries; the building of canals, and later railroads; the cheap production of cast iron and the improvement of the crude atmospheric engine into a practical prime mover, all these were the results of empirical experiments, products of craft skill and large quantities of hand labor. Only the chemical industries, from the last years of the eighteenth century, fall into a different category, for in them scientific knowledge was conspicuous. They were concerned chiefly with the preparation of the materials required for bleaching, dyeing, soapmaking, and glass manufacture. In particular, sources of natural alkalies were far from adequate to supply the demand for them, until chemical knowledge suggested processes for synthesizing alkalies from salt, limestone, and coal—readily available minerals.

By the early years of the nineteenth century the position was changing. Although most engineers and the hands who worked in chemical plants were still men who worked with their muscles and not with their heads, men employed for long hours of disagreeable, dirty, and exhausting labor, it was evident that progress in engineering demanded skill in mathematics and mechanics, just as progress in many other branches of industry now demanded chemical skill. Technological development was still dependent upon chance, empiricism, and craftsmanship (for there were as yet no machine tools) but these elements in invention were declining and those with eyes to see saw that the scientific element must increase. Many of the great entrepreneurs of the Industrial Revolution had won their fortunes by salesmanship and organizing ability, but there were examples offered by such as Watt of superior intellectual achievement. Now the Baconian claims for the usefulness of science could be reiterated with fresh vigor and support from stronger examples. To many men of science, confident that tremendous advances in material civilization were just around the corner, the wide gap that still existed between science and industry was a monstrous cause of frustration. Science should make its voice felt, and society, in its own best interests, should not only promote scientific research to hasten the millennium but submit itself willingly to the lessons that such research would teach. Curiously, about 1830, within almost every country in Europe some scientists proclaimed that the science of other countries was racing ahead under powerful encouragements, so that *their* industries must advance while *ours* stagnate. The grass was always greener over the frontier. This feeling was especially strong in Britain, which was inevitably beginning to suffer from the economic competition that first her long start in industrialization, then her isolation during the long wars (1792–1815), rendered a novel experience.

Yet the idea that society has an obligation to provide both the means of scientific instruction and those for diverting science to useful ends—these separate purposes were often combined—had been implemented long before. Though study of science in the universities was by present-day standards fantastically trivial during the early nineteenth century, it existed and was on the increase. Mathematics, then as now the essential foundation for all serious study of physical science, was taught everywhere and commonly included the applied branches (astronomy, mechanics, and a good deal of physics). Experimental training was not provided and laboratories were few and small. Linnaean botany, comparative anatomy, and zoology were working their way into the medical schools, which had long presided over chemical education, such as it was. The total effect of university teaching in science was slight, yet far from negligible, especially in the new universities, as in the United States and Prussia. Moreover, in some countries (Prussia, France, Russia) the national scientific academies offered support to a few research scientists, while the eighteenth century had seen state aid for international scientific ventures (such as the programs for the observation of the transits of Venus in 1761 and 1769) and for voyages of exploration whose function was essentially scientific. France, as befitted her role in the cultural leadership of Europe, encouraged science most. Mathematical schools had been attached to naval bases and military colleges since the seventeenth century. In the eighteenth century was founded the École des Ponts et Chaussées, which became the École Polytechnique during the Revolution. In England more was left to private enterprise. The eighteenth century was a dim period in the history of the Royal Society, but it was supplemented by the Society of Arts (1755) and the

Royal Institution (1798) just as the moribund universities were supplemented by Dissenters' Academies at Warrington, Hackney, and other centers. The two new scientific foundations were both designed to promote inventiveness for the sake of the public good. The small states of Germany and Italy had few opportunities, and with the exception of Prussia did little to exploit them. Yet some universities were very active, and trade schools came into existence here and there.

All this, together with educational reforms that stemmed from the French Revolution and Napoleon's empire, was judged far too little in the early nineteenth century, impressive as it was compared with the state of affairs even a century before. Industrialism now seemed to link technological progress and national survival. Growing populations—in Britain especially—required a constant expansion of employment (despite the fact that mechanization reduced the need for some forms of skilled labor) and, in addition, in increased food supply. Many thought that the only way to improve crop yields was to apply science to agriculture, in order to know how best to utilize different soils and to give them appropriate fertilizers. Only the United States, with apparently limitless resources of cheap virgin land, was unimpressed by this problem as yet. While the steps taken in different countries having the object of making science socially effective varied in each, they show a common pattern. One was the formation of a national Scientific Association (Germany 1823, Britain 1833, U.S.A., 1848), partly for propaganda purposes, partly to educate the public in the current progress of science. Another was the reform of university teaching, which involved the granting of degrees in scientific subjects other than mathematics and medicine, and the building of laboratories. The critical period in this respect was the two decades 1850–1870; in the United States the Morrill Land Grant Act, which made the state universities possible, was passed in 1862. In 1800 scientific research was typically conducted in the investigator's home; by 1900 it was naturalized in the university laboratory, or in that of an industrial concern. Germany led the way in many aspects of this university reform as it was applied to science, particularly in training for research; hence many American scientists studied in Germany, and imported Germanic ideas of university organization into the United States. Finally, the problem of elementary education had to be tackled, in all countries at the adult level as well as in the ordinary schools. No one yet envisaged a future in which more than a few per cent of the whole population would be college-educated, but it was obvious that prime obstacles to the introduction of science into industry and agriculture were the illiteracy and ignorance of the working people. There were many other excellent reasons for believing that real educational opportunities should be available to all; this one appealed to the profit motive as well as to idealism. Europeans could see at least this virtue in the generally higher standards and greater independence of the American people. Universal primary education was achieved in some countries, including Germany, soon after 1870, in Britain not until 1901. Meanwhile adult education, through such instruments as Mechanics' Institutes, attracted much devoted effort, including the cooperation of some scientists like Asa Gray and T. H. Huxley. Since then the growth of national wealth has permitted a large fraction of the population in a few Western countries to spend more than a third of its lifetime upon education. The rest of the world, however, is still as limited in its educational potentialities as Europe and the United States were in 1800. Recent discussions underline the fact that these problems of training future citizens for a scientific and technical society are still far from solved.

In keeping with its general philosophy of *laissez faire,* there was little public pressure on businessmen in the nineteenth century either to underwrite scientific research or to employ its results, though in England (for example) there were complaints that industrialists were less enlightened in this respect than their German rivals. In the spirit of free competition, those firms that made use of science would prosper and those that did not would collapse. Much depended on the attitude of the head of the firm, who was often the only person in it with a thorough technical grasp of the processes employed. The twin notions of (a) endowing research in the hope that it would bring forward new products or improved processes; and (b) controlling manufacture by scientific analyses of the products, were little evident before 1850 outside some sections of the chemical industry, and even at the end of the century a very great part of manufacture and even of engineering was little affected, in the day-to-day course of things, by any scientific consideration or equipment. Industrial research is always more blessed as the industrial unit grows larger; hence the United States with its huge corporations and Germany with its cartels had an advantage over the smaller firms of Britain, France, and Italy.

In the year of the Great Exhibition, 1851, Britain could still claim to be the workshop of the world. Her network of communications was incomparable. Her empire yielded cheap raw materials in vast quantity. The repeal of the corn laws (1845) permitted the free import of grain from the recently settled Middle West of the United States. Britain's mastery over the clumsy and still largely empirical technology of coal and cast iron ensured a steady market overseas for her ships, rails, locomotives, and other machinery, while Lancashire and Yorkshire dominated world textile production. Within a few years, however, there were cries of alarm. Acute observers noted the effects of American mass production and the system of manufacture with interchangeable parts whose invention is traditionally assigned to Eli Whitney. In the United States engineering was making profitable use of advanced types of machine tool, such as automatic lathes and gear cutters, hardly known across the Atlantic. The Civil War, with its enormous demand for munitions, transport, and clothing, had stimulated a rapid and permanent growth of the industry of the northeastern states, which heavy immigration after the war maintained. American invention turned steadily towards increasing the quantity of production, and decreasing the need for skilled labor. It resulted in the McCormick reaper, the sewing machine, the steel converter, the elevator, and countless other devices now taken for granted. After the Civil War, too, the United States led the way in new industries, particularly oil and electric power. Thomas Alva Edison (1847–1931) made invention a business. By 1900 the United States had emerged as one of the world's industrial giants. Germany was the other. The unification of Germany (1871) and the achievement of her political hegemony on the Continent had provided the final conditions required for the booming growth of the industries that had developed steadily in the Rhineland basin, in Saxony, and the North Sea ports.

Steel was now the most important of all materials. The automobile was in production, and flight about to be accomplished. Radio telegraphy had crossed the Atlantic, skyscrapers had been built, and electric trams and trains were running. It was possible to telephone, to read by electric light, to have one's appendix removed under an anesthetic in a sterile operating room. The machine gun was ready for war. Canning and cold storage preserved foodstuffs. Opinion was formed by mass-circulation newspapers rushing off steam-driven rotary presses. The twentieth-century world was assuming its shape.

To what extent must science be praised, or blamed, for this new creation? At the present time no confident answer is possible, for we know all too little about the development of nineteenth-century technology. It is easy to discern that some innovations would have been impossible without science: the medical advances of Pasteur, Lister, and Koch; the manufacture of aniline dyes, of cellulose products, of many drugs which derived from knowledge of organic chemistry; the electrical industry based on the discoveries of Oersted and Faraday. Yet, even in these fields, the fundamental principles were commonly far from clear, so that chemists, for example, had to do much experimental fumbling in the dark before they hit upon some new substance that was advantageous as a dye or a drug. Similarly in the electrical industry a great deal of "technical development"—empirical engineering—took place before satisfactory generators, motors, incandescent lamps, and so forth, were made. The best design for none of these things could be deduced from existing theory. As a result the man who foresaw practical applications for some effect known to science, and developed it to usefulness, was a figure of particular significance: such men as Edison, Bell, Swan, Siemens, Steinmetz, Marconi, and the Wright brothers, whose success sprang from their careful study of aerodynamics. None of these enunciated important new scientific truths, yet by no means did they borrow their discoveries out of scientific textbooks. Their success was brought about by scientific examination, with an eye to practical utility rather than to the increase of knowledge, of matters which science had noted but passed by. In this respect, then, it may be said that science had become, through the nineteenth century, potentially far more serviceable to technology than it had ever been in the past, if only because the range of knowledge and the breadth of the effects upon which science spoke was now so vast. Yet still science offered very few ready-made solutions.

On the other hand there were industries like mining and metal smelting in which the conditions of labor, though improved, were still extremely harsh. Though powerful machinery had diminished muscular effort and increased production, the principles of such industries remained medieval. Again, the mechanical inventiveness that produced the sewing machine or the gas engine was no different in kind from that which had produced the first mechanical clocks and water mills, though of course methods of fabrication were very different. The chief scientific principle involved—if such it can be called—was the making of parts to closely measured tolerances with the aid of elaborate machine tools, gauges, and micrometers. But there was no more science built into a bicycle than there was into a stone ax.

It is important to recognize that technology undergoes its own conceptual development independently of science. Something like a modern automobile could have been constructed in 1900, if the idea of it had existed; instead, with great pride and consciousness of progress, men manufactured a horseless carriage. Two generations earlier they had placed stage coaches on railway lines, and cast Gothic capitals in iron; their first steamships were square-riggers with an awkward funnel, and so on. The evolution from the rudimentary specimen to the finished form, as it seems to us, of an automobile, a steamship, or a table knife is affected surprisingly little by scientific progress and to a very great extent by engineering development and experience in design. Or rather, this was still the case until the twentieth century was well advanced.

To that extent the impact of science upon technology, and through technology upon society, was limited throughout the nineteenth century. The Baconian spirit was triumphant, but science was indubitably the servant of man—the servant of classically educated politicians, of cautious businessmen and benevolently bearded family doctors. No one yet thought of the scientist as a Frankenstein, or seriously doubted society's ability to control and order whatever powers he might reveal. Of course everyone in the later nineteenth century recognized that the world had changed greatly, and was continuing to change swiftly, and science was seen to have played its part in this course of change—though industrial and political developments were generally given first place as creators of change. Most people complacently identified change with progress; as they spoke for the European races, who can blame them? Yet no one, not even Jules Verne and Wells, foresaw the magnitude of the second scientific revolution that lay little more than a generation ahead. What impresses the contemporary reader of nineteenth-century science fiction is not the submarine or the flying machines, but the archaic character of the world into which these devices were artificially inserted. It was not very difficult, two generations ago, to imagine that science would accomplish flight or even a journey to the moon; what could not be foreseen was the utter disappearance of the Victorian world, and with that the final rupture of so many historic threads that had given European and American society its consistent, homogeneous character.

Chapter 13

NATURE, TIME, AND MAN

It is difficult to imagine any more inherently fascinating subject for speculation than how the universe came to be as it is. "Since I was first inclin'd to the Contemplation of Nature, and took Pleasure to trace out the Causes of Effects and the Dependance of one thing upon another in the visible Creation, I had always, methought, a particular Curiosity to look back into the Sources and Originals of Things,"[1] wrote Thomas Burnet in 1681, and his contemporaries showed the same interest. Though the Bible provided what Christians were commanded to accept as a definitive account of the creation of the solar system, the Earth, plants, animals, and man, by the eighteenth century scientists were in grave disagreement with the traditional view. Theologians, naturalists, and astronomers combined to suggest a more complex development than that provided in *Genesis,* and scientific fact and theory combined to make evolution the most influential concept of nineteenth-century thought.

Concrete speculations on evolutionary geology and cosmology long preceded any on biological development. The intensely rational attitude of the seventeenth century militated against the medieval view of fossils as "thunderstones," or the Renaissance view that they were aspects of God's overabundant creative power; they were rather taken to be clear indications that seas had once existed where now there was dry land. Those who looked saw evidence of aqueous erosion, comparable to, but vastly greater than, what could be observed as day-to-day phenomena. What more reasonable in a devout age than to assume that the agency was Noah's Flood? But even this was not quite reasonable; for the forty days and forty nights could not have produced a sufficient depth of water to cover all the mountains. So argued Burnet, insisting moreover that God would never have created such imperfect and hideous shapes. (The Romantic age with its taste for wild scenery was still a century ahead.) Burnet suggested ingeniously that the Earth must have been created as a perfect sphere; it was the Flood which had raised mountains, uncovered rocks and deposited fossils. Greatly daring, the heterodox Whiston in 1696 suggested that the chaos out of which God created the Earth was the atmosphere of a passing comet. Half a century later the naturalist Buffon could go no further than to multiply the traditional time scale (five to six millennia) by about ten times and to break down the development of the Earth, after it had been formed as a lump of matter knocked out of the Sun by a passing comet, into a discrete series of steps.

Buffon's theory was too speculative for the scientists, too radical for the theologians. The first scientifically respectable theory of cosmological evolution was the nebular hypothesis which Laplace tentatively advanced in an appendix to his *System of the World* (1796), at a time when there was no need to placate theologians. This hypothesis, soon advanced by others to the rank of accepted scientific theory, had the great merit of explaining the most notable physical symmetries, such as the orientation of the orbits and directions of rotation of the planets and their satellites. It was a theory highly acceptable to astronomers, but little known to laymen.

The idea of geological evolution, meanwhile, was experiencing some very peculiar vicissitudes. The eighteenth century's passion for taxonomy, so strongly revealed in botanical studies, is apparent in the attempts to classify rocks. The center of this work was in the Mining Schools of Germany; here professors like J. G. Lehmann (d. 1767) lectured indifferently upon the relation of ore-bearing veins to the "golden tree" with its roots in the center of the Earth and upon the apparently orderly manner in which the Earth's crust was composed of layers or strata lying in a definite order, varying with the age of the mountain ("primitive" dating from the creation, "secondary" created by the Flood, and "tertiary" by natural action since the Flood) of which they form a part.*

Gradually this mysticism was transformed into a generalizing theory, partly based upon mineralogical study and classification, partly upon a transcendent belief in the harmony and unity of nature. A. G. Werner (1749–1817) at Freiburg, in the manner common in German universities in the nineteenth century, developed a "school" whose tenets were spread far and wide by his pupils. Werner's theory (soon to be known as the Neptunian theory) was that a universal ocean (to be identified with the Flood by the pious) originally covered what is now the core of the Earth; this ocean held in suspension a vast variety of particles whose gradual precipitation caused the observed rock strata to build up. Finally the ocean vanished (Werner never bothered to establish a precise mechanism) leaving the Earth as we now know it. Though Werner recognized that volcanoes existed he vehemently denied that there could be any trace of volcanic action in northern Europe, or that there could be any heat or fire below the surface of the Earth. He supported his theory with a host of observations drawn from surveys of central Europe; he conveniently ignored or denied evidences of volcanic action detected in France by field geologists like Guettard (1715–1786) and Desmarest (1725–1815) who had correctly determined the connection between lava and basaltic rocks.

By 1790 the Wernerian or Neptunian theory was sufficiently well established to evoke an opposition theory, the Plutonian or Volcanic theory. The Plutonian theory was never so dogmatic or so personal as the Neptunian, except perhaps in the Scottish universities where the partisan spirit was rife. The Plutonian theory is mainly associated with the name of James Hutton (1726–1796), a gentleman farmer and chemical industrialist. Hutton's publications, and still more the work of his friend and disciple John Playfair, presented a view of the Earth's development which differed remarkably from that of Werner in both content and method. Long observation

* The most remarkable example of the vagaries of eighteenth-century professors is to be found in a book published in 1726 by Johann Beringer, professor at the University of Würzburg. Convinced that fossils were formed by God as signs and symbols, he was delighted to find examples covered with Hebrew characters, others with strange creatures and even astronomical bodies, and triumphantly pictured them in a large book. Only discovery of a "fossil" bearing his own name showed him his error: in fact all his most interesting specimens had been faked by his students, far more enlightened on this subject than their professor.

had convinced Hutton that many surface rocks were extrusions from below, and showed definite traces of volcanic origin. His friend Sir James Hall, inventing thereby the very idea of experimental geology, performed a number of experiments on cooling molten glass; he found that while quick cooling produced the familiar glasslike material, slow cooling produced a crystalline substance. Hall and Hutton applied this discovery to explain how granite could be formed by volcanic action alone. At the same time Hutton not only recognized the existence of sedimentary rocks (formed by the deposition of solid matter from an aqueous solution); he insisted that the processes which had in the past formed the Earth were similar to those which continued in his own day.

This last principle, christened *uniformitarianism* by Charles Lyell (1795–1875), was a powerful weapon for common sense, and did much to quiet the alarm of pious people who felt that the extreme Neptunians and Plutonians, the Catastrophists, were trying to upset orthodox belief. The idea of uniformitarianism was scientifically more acceptable as well. Together with a wealth of data collected by field geologists on a variety of subjects—mineralogy, crystal structure, stratification, geological surveys, fossil stratification, and paleontology—it forms the basis for Lyell's *Principles of Geology* (1830–1833), which, more than any other work, established evolutionary geology, at least in the English-speaking world, on a sound and viable basis. Geology was henceforth a self-confident, specialized science.

Concepts of biological evolution developed less from empiricism and the rationalization of the Biblical account than from philosophical notions of the underlying unity of life. The very notion of taxonomy as developed in the eighteenth century implied unity while stressing fixity. There was clearly an underlying, unifying plan in God's creation: this was equally expressed in the Great Chain of Being (the idea that all living things, from the lowest to the highest animal—man—were connected in a stepwise development that contained no temporal element), with its belief in the plenitude of nature, and in the invention of the word biology (simultaneously in 1800 by a German and a Frenchman). The same philosophical presuppositions which later made physicists seek to correlate physical forces made botanists and zoologists eager to find principles to guide their study of living matter. In all this, man was taken as the highest of animals, marked off by the possession of a rational soul, but purely animal in body.

As long as there was no question of *evolution,* the animal nature of man—accepted since antiquity—caused no emotional or religious pangs. But the growth of evolutionary geology inevitably suggests first (as for Buffon) that lower animals appeared on the Earth before the higher ones—perhaps merely through God's creation—and then, more daringly, that the Great Chain of Being might be temporalized to produce true evolution, the development of higher forms from lower ones.

Empirical evidence for any such development was at first exceedingly meager, and long remained so. There was the evidence of paleontology, for what it was worth: that various plants, animals, and (especially) forms of marine life had existed in the remote past, and now did so no longer: this showed that the past differed from the present, but did not require an evolutionary explanation. There was the evidence of taxonomy: it was increasingly difficult to classify species without overlapping; against this was the fact that species did in fact exist. There was the evidence of man's effect upon the environment: the various changes wrought by man in domestic animals, from dogs to cattle, and in food plants; but this was not a fact noticed by those who were not predisposed to an evolutionary idea. The earliest examples of evolutionary schemes are marked by a notable absence of empirical data; a marked mysticism, whether in the form of *Naturphilosophie* or simple piety; and, as in the case of geology, a strong strain of rationalizing theology. So Erasmus Darwin (1731–1802) in *Zoonomia or the Laws of Organic Life* (1794) imagined as an interesting speculation that

> all warm-blooded animals have arisen from one living filament, which the great First Cause endued with animality, with the power of acquiring new parts, attended with new propensities, directed by irritations, sensations, volitions, and associations, and thus possessing the faculty of continuing to improve by its own inherent activity, and of delivering down these improvements by generation to its posterity, world without end.[2]

In such dithyrambic prose Darwin described, with surprising attention to empirical detail, to plant hybridization, to the domestication of animals, and to the spontaneous appearance of anomalous forms, a somewhat vague but definite theory of evolution. He was but one of many to do so: the most famous English example is *Vestiges of the Natural History of Creation* (1844), an extremely successful work of popularization which professed to survey "the whole realm" of nature, all from an evolutionary point of view. Such works made the popular mind familiar with evolutionary notions; they are not in themselves scientific contributions.

Only one writer before Darwin deserves to be taken seriously, because only he attempted to explain the mechanism by which one species might develop out of another. This is J. B. Lamarck (1744–1829), successively an obscure botanist and the creator of invertebrate zoology. Lamarck viewed life as a series of "organic movements" and it was these which, as he saw it, were responsible for the gradual development of creatures into the form with which we are familiar. In his *Zoological Philosophy* (1809) Lamarck discussed both evolution and the nature of life. In his view all living things had evolved from vague low forms (the products of spontaneous generation) by reaction to environmentally caused needs. Thus a swamp plant might respond to the drying up of its environment either by dying, or by altering its physiology in such a way as to require less water; its offspring, inheriting this variation, might further respond until a dry-land species was evolved. In the famous case of the giraffe, Lamarck postulated a change in environment which would make the eating of leaves a necessity for survival; to this an individual might respond by trying to reach ever higher leaves, and in so doing stimulate the organic fluids which, in response, would cause the neck to grow—successive generations creating the familiar and improbable creature we know as a giraffe. Lamarck's great weakness lay in this concept of organic fluids; educated in the physics of the eighteenth century and utilizing chemical concepts untouched by the work of Lavoisier, his mechanism was clumsy, obscure, and open to misinterpretation. No wonder his critics jeered at the man who thought a creature developed an organ by intent and will, or that few French natural scientists—Geoffroy St. Hilaire was a unique exception in the Academy of Science—took his work seriously. His failure to devise a credible mechanism classified Lamarck as a woolly nature-philosopher and vitalist, and the sound features of his system were obscured. It continued to remain the case that, while ideas of geological evolution grew ever more firmly a part of science, ideas of biological evolution belonged either to the mystic or to the nonscientist.

It is hardly surprising, then, that Charles Darwin (1809–1882), who was to devise the first complete and scientifically respectable theory of evolution, should have begun his scientific career strongly prejudiced against the evolutionary idea.

His views changed during the years spent aboard the ship *Beagle* in and around the coasts of South America (1831–1836) in the course of which he read Lyell's *Principles of Geology,* observed the ecology of the pampas, and visited the Galápagos Islands. Upon his return, convinced of the *reality* of evolution, he began collecting facts which should eventually, he hoped, lead him to an understanding of these strange phenomena—perhaps the only perfect example of Baconian method in the whole history of science. The mechanism still escaped him until, as he said, "happening" to read Malthus for pleasure, he saw that the Malthusian doctrine of the struggle by an ever-expanding population for a more slowly expanding food supply offered the key.

It was the tendency of all organisms to increase in geometrical ratio, while the population of any one species tends to remain constant, that led Darwin to the first important factor of the mechanism of evolution, the "struggle for existence" manifested so widely in the natural world. Yet this alone did not explain change; one more fact must be noted: that all organisms vary appreciably. Darwin recollected his South American experience, when he had observed during a drought on the pampas that ordinary cattle survived, while a much-prized short-jawed breed starved; he concluded that, as in this case, when there is competition for the food supply favorable variations will permit survival, and this variation, if transmitted to the offspring, will slowly change the species. Any variation which helped an individual to survive—whether by getting more food, evading its enemies, adapting to a changing environment or gaining a mate *—would increase the chance of that variation's being preserved by posterity. Darwin stressed that all variation was random, both favorable and unfavorable variations occurring with equal probability; that all variation was slight; and that no agency was involved in causing it to occur. He used the term "natural selection" to describe how variations were perpetuated, but he thought of this as a passive force. Nature did not consciously select, like a dog breeder, but unfavorable variations made it harder for the individual to survive, favorable ones easier. This was not a case of the "survival of the fittest," a term coined by the philosopher–sociologist Herbert Spencer

* Darwin at first found it impossible to understand how a peacock's beautiful but cumbersome tail could assist it in food gathering. When the idea of sexual selection occurred to him the problem was solved.

and repudiated by Darwin, who equated *favorable* with *most useful*. This theory of natural selection permitted the understanding of many curious facts hitherto explained as examples of the effect of use and disuse: e.g., for Darwin the blind fish in deep caves had not lost their ability to see through disuse; rather a natural variation—total or partial blindness—was no disadvantage in this environment, and so was not eliminated. Darwin was of course wrong in assuming that all variations were slight; and (like all evolutionists) he was forced to allow the inheritance of new characteristics, though at least his inheritable characteristics were those revealed at birth, not acquired during life. There was still the fact that in most cases gross defects are not inherited, and there was as yet no means whereby to determine what kinds of variations were transmissible. But these were not the main objections voiced by Darwin's contemporaries; they spoke rather of the imperfection of the geological record and the speculativeness of the whole idea.

In fact, *The Origin of Species* (1859) * is not a speculative work, nor is it in the class of popular science. It is a serious, factual, detailed, closely knit work on biology, of a kind never previously attempted on the subject of evolution. Darwin drew on an immense collection of facts (and, in the case of plants, his own experiments) in describing "domestic" selection, an even wider range of facts on "natural" selection; described his theory; squarely faced its difficulties, including the imperfection of the geologic record; utilized a vast survey of geologic distribution to show how organisms adapted to various environments, and, for good measure, included an immense amount of detail on the anatomy, morphology, and habits of a most varied lot of creatures. Here he drew on his own great botanical and zoological knowledge, on the work of his contemporaries, and the help of his friends, from Huxley in England to Asa Gray in America. It was meant to convince, and it did convince all except those who, for one reason or another, found its doctrine utterly distasteful.

These latter were all too often those who objected to the

* Darwin planned an enormous, multivolumed work. In 1844 he wrote a sketch, with an outline of the longer work, for circulation to his friends. The great work was never written; in 1858 an unknown young naturalist, A. R. Wallace, sent Darwin a paper which contained the theory of natural selection. Darwin generously intended to see that this was published without comment; Darwin's friends forced him first to submit a paper to the Linnaean Society along with Wallace's, and then to write a full-length book.

whole notion of progressive development because, as Asa Gray (who confessed to finding all new ideas "annoying") put it in his review of Darwin's book in the *Atlantic Monthly* for 1860, "The very first step backward makes the Negro and the Hottentot our blood relation—not that reason or Scripture objects to that, though pride may",[3] yet for him scientific acumen and honesty overcame distaste, and he became a convinced Darwinian. His Harvard confrère, the Swiss Louis Agassiz, objected first on the grounds of the imperfection of the geological record; and secondly (and more strongly) on the grounds that species represented God's plan and that Darwin's theory therefore interfered with the argument from design. (To which Gray wryly retorted that Agassiz's faith was very weak, if biological facts could shake it.) Darwin was attacked, by Catholics and Protestants alike, as an infidel atheist making "the ape our Adam" and declaring that the Bible was a monstrous lie. Darwin replied mildly that it increased God's grandeur and wonder to believe that the universe had been created with evolution built in; in any case he had not discussed man (though he was soon to do so). The anti-Darwinians nearly all responded as Asa Gray had done; that descent from lower forms made man too nearly akin to a monkey, a somewhat curious reaction since, at least until 1800, the similarities of structure between man and animals had been taken for granted. More serious criticism came on the subject of variation.

That before Darwin died his *Origin of Species* had been reprinted many times, and its thesis widely accepted, was a tribute to its importance as well as to the fact that most intelligent theologians realized (a trifle belatedly) that to preach against accepted scientific doctrine was to cast doubt upon theology, and incur ridicule. With acceptance, Darwin's theory began to change profoundly the content of biology. On the one hand, natural history turned into ecology, the study of the structure of animal and plant populations and their relation to their environments *—more interesting and more useful to science than forming collections of plants or studying animal habits. On the other hand, of far more fundamental importance, Darwin's evolutionary theory en-

* So Darwin once (not very seriously) commented upon the direct connection between the number of old maids in a village and the luxuriance of red clover in the surrounding fields. Red clover is fertilized by only one kind of bee, whose nests are destroyed by field mice; cats eat mice; and old maids keep cats. Therefore . . .

couraged, indeed demanded, a study of variation and its transmission. He had no theory of how or why variation occurred, and no notion of when it might be predictably transferable. Both pro- and antievolutionists saw this as a critical problem essential to the proper testing of Darwin's theory.

The simplest view was that each individual is the sum of his ancestors; Darwin's cousin Francis Galton (1822–1911) thought he had demonstrated this both from statistical considerations and from a survey of "gifted" families like his own. His smug assertion that genius runs in families and environmental effects are negligible led to the dubious speculations associated with eugenics (his term). On the other hand his interest in statistics was turned into a powerful scientific tool by Karl Pearson (1857–1936) and others, and has become indispensable for modern genetics.

A truly scientific study of heredity was possible only after the role of the cell was better understood. Weismann's theory of the germ plasm as the conveyor of "hereditary elements" (cf. p. 247) seems to have suggested to Hugo de Vries (1848–1935) that inheritance might proceed by the transmission of particles. About 1886 de Vries observed interesting variations in the American evening primrose; study showed him that these plants produced discontinuous variations of exactly the kind he was looking for, and these large variations he called *mutations*. A few years later William Bateson (1861–1926) independently found much the same thing. Slowly biologists began to regard this work as important, and recollected that there had been earlier work of the same kind. In 1900, de Vries and others sought for such work and rediscovered that of Gregor Mendel (1822–1884) first published in the 1860s but previously somewhat neglected because it appeared to be inexplicable, and was certainly unrelated to biological problems then current.

Mendel, endeavoring to understand heredity, worked with sweet peas, normally self-fertilized. The most obvious distinction in characteristics was between a tall and a short variety; when these were crossed, their offspring, according to the theory of Galton and others, should have ranged from short to tall with many intermediate heights. But in the event all were tall. This was sufficiently odd, but even odder was that these tall offspring were not identical, for when they in turn produced offspring by self-fertilization one-quarter were short and bred true, one-quarter were tall and bred true, and one-half were tall and in turn bred one-quarter short, and so on. Or, as it is more usually represented

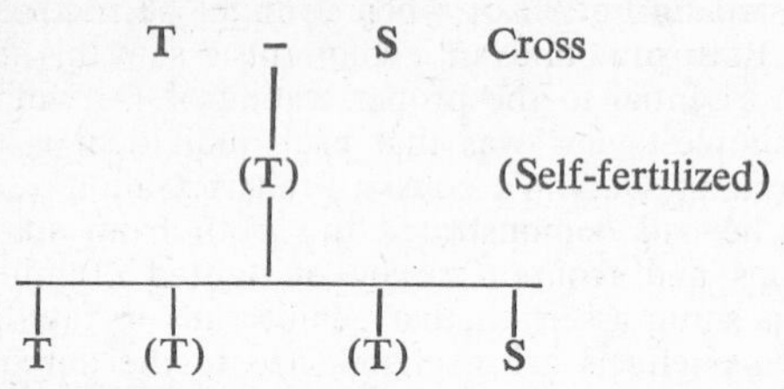

In this case Mendel called tallness a *dominant* characteristic and shortness a *recessive* characteristic, and concluded that there must be two factors for height present in each individual. So in the case of (T) peas, there are both T and S factors; while those which breed true have either T or S factors alone. Since tallness is dominant, an individual with both factors will always be tall.

The Mendelian theory of inheritance, combined with the discovery of mutation and associated with investigations of the chromosomes (see pp. 247–248), offered a possible extension of Darwin's theory. The "hereditary factors" of de Vries were christened "genes" in 1909, and cytologists extended the study of chromosome development. But there were difficulties as well and Mendelian heredity was not universally accepted. It was soon discovered, by T. H. Morgan (1866–1945) and others, that many characteristics did not behave like Mendelian factors; these were later shown to be "linked" to other characteristics, most often to a particular sex. Other work showed that variations must be distinguished into two main varieties: those inherent in the individual which are not amenable to external influences (genotypic), and those (usually more obvious) variations which are influenced by environment and not inherited (phenotypic). The situation was further complicated by the discovery of H. J. Muller in 1927 that mutations can be produced by X rays; these mutations are generally of a harmful or useless type (as indeed are most natural mutations) but their artificial production, especially in the quickly reproducing fruit fly, *Drosophila melanogaster*, facilitates the observation of the complex relation between heredity and mutations. Subsequent work, especially by R. A. Fisher, J. B. S. Haldane, and T. Dobzhansky has concentrated upon detailed statistical analysis of variations in genetic combination as well as variations in population, all tending to support the role of natural selection in genetic heredity. Though the relative importance of heredity and environment has been variously assessed in different periods, modern genetics relies upon modified Darwinian and Mendelian considerations.

Almost no one could read *The Origin of Species* without wondering how natural selection applied to man. The very violence of the popular and theological reaction would, in any case, have forced zoologists to try to locate man in the evolutionary scale. Lyell, T. H. Huxley (1825–1895), and Darwin all tried; Huxley's *Man's Place in Nature* (1863) is particularly important for its intensive study of comparative anatomy and its conclusion that man was physically closer to the higher apes than these were to the monkeys. Huxley also opened a new field of study by insisting that the Neanderthal skull (discovered in 1856) was a genuinely ancient skull, probably that of an earlier human species, and not that of a modern deformed idiot. Though the subsequent search for "missing links" was responsible for the altogether confusing Piltdown hoax, it was equally responsible for bring nascent archeology to the attention of zoologists and for helping to establish it as a respectable study.

Prehistoric archeology had begun casually enough, as the offshoot of antiquarianism coupled with an esthetic interest in nature. To collect fossil remains or chipped flints demanded no specific knowledge or theory; in time, as collecting grew, it stimulated the desire to understand the purpose of these strange artifacts. In the late seventeenth and early eighteenth centuries they were associated with the Druids, as were the megalithic circles of Brittany and England. Cook's discovery of Stone Age cultures in the Pacific Ocean suggested that flint implements were the relic of stone-using men necessarily far older than the Druids. By the early nineteenth century they were widely though not universally accepted as instruments, and advances in geology permitted some realization of their great antiquity. Systematic investigations were undertaken: in 1837 Jacques Boucher de Perthes (1788–1867) began a ten-year search of the gravel beds of the Somme, having detected the stratifications caused by the river as it cut its bed deeper and deeper. Though at first he thought his "axes" were Celtic or "diluvian," he soon realized that those found in the same beds as extinct animals must be "antediluvian." His enthusiasm was not enough to convince those like the zoologist Cuvier who saw the world as having been created in a short time by a series of catastrophes, and he never was able to find any direct connection between man and the implements.* He received some credence in England, especially after 1858, where similar work was in progress; those familiar with the evidence were convinced that stone implements were made by men living at the time when prehistoric animals flourished. Soon, French archeologists were recognizing two main types of worked stones; in 1865 Lubbock (1834–1913) coined the terms "paleolithic" and "neolithic" to describe the two quite different cultures and ages. As the Bronze Age cultures of Egypt, Mesopotamia, and Greece were systematically explored and some idea of their antiquity achieved, it became clear that the Stone Age cultures must lie far back in the past. At the same time the development of cultural anthropology (ethnology) showed that this was a universal phase in man's development. More precise geological knowledge permitted a clearer picture of the time span involved, and man's history began to stretch back into the remote past, now beginning to emerge clearly.

Contemporary with advances in prehistoric archeology was the development of ethnology, with special emphasis on the evolution of man from a (presumed) animal stage to "civilized" man. Cultural anthropology in the nineteenth century had its roots in philological studies, coupled with a new interest in myth and folk legend expressed in the collection of fairy tales, regarded as relics of "folk memory." Biological evolution, the intensification of historical studies, and the opening up of hitherto unknown areas, like Mexico and the South Seas, made possible the careful study of primitive culture. For example E. B. Tylor (1832–1917), though not himself a field anthropologist, had much to do with establishing criteria of reliability for data gathered by others. He particularly stressed the study of institutions and of religion, insisting on the evolutionary nature of all aspects of human society. Gradually the fact that human society had evolved through various recognizable stages produced a series of classifications, mostly based upon expanding technology.

One very dangerous aspect of cultural anthropology, also intimately associated with evolutionary concepts, was the outburst of racism. Darwinian natural selection, carelessly as-

* Boucher de Perthes thought he had found skeletal remains and axes together in 1863, but the jawbone and many of the implements proved to be forgeries made by workmen, anxious to collect the previously announced reward. It was only later that authentic finds of graves containing worked flints were made.

sociated with Spencer's "survival of the fittest," led to a wide variety of aberrations, from "social Darwinism" to "might makes right." By an easy logic, "fittest" came to be interpreted as "best," and cutthroat competition, whether between nations or between businessmen, to be complacently justified and even glorified as the instrument whereby social evolution could prevent the survival of the unfit. Though Darwin protested that this was not what he had meant, and Huxley tried to insist that civilized society existed to prevent the merely strong from being the sole survivors, social Darwinism was too convenient a justification for imperialism, war, free enterprise, and avoidance of social legislation to die easily. Even the new science of sociology, derived from Auguste Comte's view that a science of man was a necessity as soon as the sciences of inanimate and animate nature were established, for a time succumbed to the belief that nature must be allowed to take its course. Just as the eighteenth century had tried to apply the laws of physical science to society, the later nineteenth tried to apply the laws of biological science to justify its view of what society ought to be. By 1900 the mechanics of Darwinian evolution had been radically revised by geneticists; they were still valid for the new social sciences, which awaited the twentieth century to reach maturity.

Chapter 14

THE SCIENCE OF LIFE

The ineffectiveness of biological thought from the late eighteenth into the early nineteenth century is most cruelly revealed in medical theory, reflected (unfortunately) in the physician's practice. One stream of ideas, the oldest and still perhaps the strongest, descended straight from Galen: it taught that disease was a disturbance of the balance of the four bodily humors caused in turn by an environment, a diet, or a manner of life that was unhealthy. It insisted on purgation and bloodletting as the strongholds of treatment. Other philosophies of medicine were both more recent and more simple. Thus the Scottish physician John Brown (1735–1788) divided all diseases into two forms, involving either hypertension or undue relaxation of muscle and nerve. His American contemporary Rush declared that *all* diseases resulted from hypertension. Only a little later a French professor of medicine, F. J. V. Broussais (1772–1838), no less dogmatically assigned all sickness to irritation of the intestinal tract. It is hardly surprising that, after this, the so-called "therapeutic nihilists" professed openly the physician's inability to cure the sick. Systems were foolish, drugs useless.

Despite frequent pretensions to the contrary, prescient suggestions, and some truly remarkable early successes (such as vaccination), scientific medicine only came into being in the second half of the nineteenth century. Since its twin foundations were experimental physiology and microbiology, the year 1865, in which Claude Bernard (1813–1878) published his book on *Experimental Medicine* and Louis Pasteur (1822–1895) began his investigation of the silkworm disease, stands out as an obvious landmark. In this year also Lister initiated antiseptic surgery; anesthetics, the stethoscope, and the clinical thermometer were all of recent introduction. But the greatest of the innovations was the conception of the medical man as an applied scientist, the social cutting edge of the tools supplied by the bacteriologist, the organic chemist, the cytologist, the physicist, the physiologist, and so on. This was a change that could only come about because these sciences had developed rapidly enough by 1865 or soon thereafter to make their application to medicine successful.

Although knowledge of the distribution, variety, and even history of living things improved rapidly in the period from about 1760 to 1820 (Chapter 13), knowledge of living *processes* remained stagnant. Something was learned from the new chemistry of Lavoisier's successors, but the application of the microscope and the development of physiological experiment were yet to come. In the early nineteenth century physiology was still, for Bichat in Paris, as it had been for the Hunters in London and the Monros in Edinburgh, the offspring of anatomy. The great physiological discovery of this period was a direct result of anatomical work on living specimens, in the tradition of Galen and Vesalius. Credit for the discrimination between the motor function of the anterior spinal nerve roots and the sensory function of the posterior roots, which laid the basis for the modern study of nervous physiology, is properly divided between an Englishman, Charles Bell (1774–1842), and his French contemporary, François Magendie.

Unlike Bell, who remained essentially an anatomist, Magendie, having begun his career in the same way, was led in a different direction. Beginning with an important series of experiments upon the effects of various drugs upon animals, at a time when experimental pharmacology was in its infancy, Magendie later extended the techniques of experimenting upon living animals to the investigation of such physiological questions as the nature of the mechanism of swallowing and regurgitation, of digestion, and of the circulation. He also undertook some purely medical researches of the same kind, proving for instance that a dog became rabid when the saliva of a patient suffering from hydrophobia was injected into it. The method of animal experiment was further refined by Magendie's greater pupil, Bernard. In combining it with a careful and imaginative use of chemical analysis Bernard devised one of the physiologist's most powerful tools. The classic instance of his own use of it is the discovery (1848–1857) of the glycogenic function of the liver, by which carbohydrate is stored for conversion into sugar.

Making the then universal assumption that sugar in the blood served as a fuel in the respiratory metabolism, and that all this sugar derived from the diet, Bernard wished to trace the path of the sugar more completely. What followed he attributed entirely to his systematic pursuit of the empirical method. When a dog was fed on food containing sugar he found, as he expected, sugar in the superhepatic veins: this was revealed by a chemical test. Then Bernard made what he called the "comparative experiment": another dog was fed on a meat diet free from sugar and carbohydrate, so that in its veins Bernard expected to find no sugar. He was wrong, and learned through further experiments that the blood of all animals, even when fasting, contains sugar. Its origin he traced to the liver; an excised liver washed clean of sugar actually produced more after standing for some hours. This resulted from the action of an enzyme upon a starchlike substance which Bernard isolated from the liver and called *glycogen* (sugar-former). When an animal is well nourished, glycogen is stored in the liver; as the level of blood sugar falls after a meal it is maintained by the conversion of glycogen into sugar. Thus Bernard demonstrated, for the first time, the synthesis of complex substances in the animal.

Just as Bell's work followed the British tradition of "experimental anatomy," and Magendie's the French, while Bernard (and to some extent Pasteur) carried forward the researches on animals initiated by Magendie, so there was another national tradition of physiology in Germany that exhibited at first the same parallelism to what occurred in France and Britain. Johannes Müller (1801–1858), who taught at Berlin and influenced a whole generation of German biologists,* came to physiology from zoology. He too worked on the nervous system, and is always associated with the discovery of the "law of specific nerve energies" (1840)—each sensory organ responds to any stimulus with its own particular sensation. Müller's physiologist pupils, however though they learned much from their master, joined their

* Among them Theodor Schwann (1810–1882), E. DuBois-Reymond (1818–1896), H. L. F. von Helmholtz (1821–1894), and Rudolf Virchow (1821–1902).

French colleagues in reacting against the vitalism that Müller espoused. Unlike the French, they performed many experiments on nerve–muscle preparations (mainly from frogs): DuBois-Reymond's drastically mechanistic study of animal electricity culminated in his discovery that muscles in the active state produce measurable electrical impulses; Helmholtz succeeded in actually measuring the time interval between nervous stimulus and muscular reaction—which Müller had said could never be measured.

While physiology was contending, in the first half of the nineteenth century, with the formulation of questions that experiment could answer, as well as with the development of appropriate techniques for getting the answers, all biological activity was penetrated by a profound philosophical issue: that of mechanism against vitalism. Every physiologist was forced by the nature of his work to take sides on this question. In one guise or another, vitalism—the notion that the phenomena associated with living things are unique in kind, and only reproducible by living agencies *—prevailed everywhere in the early part of the century. In England it was represented by William Paley and the authors of the *Bridgewater Treatises* (Bell among them) with their emphasis on design; in France by the philosopher–anatomist M. F. X. Bichat (1771–1802); but the most extreme and pervasive version of vitalism was that purveyed by the German *Naturphilosophen,* making much of life-forces, organic molecules, and so forth. The trend of the nineteenth century was away from vitalism, however, and—in physiology—towards the belief that living phenomena would ultimately be fully explained in chemical and physical terms. Limited as the competence of such "mechanistic" explanations was about 1850, their philosophical import was strengthened by the growth of evolutionary ideas. Mystic, Biblically inspired hypotheses were everywhere under question.

Bichat had especially emphasized the "*instability* of the vital forces [that marked] all vital phenomena with an *irregularity* which distinguishes them from physical phenomena, remarkable for their uniformity."[1] That is, the physicist's world is law-governed; the biologist's is not. To take this literally would be to render meaningless every endeavor towards a science of biology. Indeed, medicine itself would become hopeless if there were no consistency in the

* One French physiologist, inquiring how nitrogenous compounds (proteins) are formed by vegetarian animals, suggested that organic nitrogen could be created by nonchemical, vital processes.

course of disease, or in patients' reactions to treatment Bichat's principle was peculiarly destructive of physiology, biological science having neither description nor classificatio as its objective, but *explanation*. For the power to explai implies regularity of cause and effect, not to say consistency The budding physiologist Magendie attacked Bichat's concep as early as 1809, insisting that even a vital force, if it exists must be consistent in its operation. This was the positio that Bernard was to emphasize and strengthen, especially i his book on experimental medicine.

For Bernard was, besides all else, a philosopher of science and a firm adherent to an experimental method of research and reasoning that he believed could be defined, illustrated, and followed by others as successfully as by himself. He clearly saw, too, how his trust in this method was inextricably linked to his philosophical position as a "determinist." The determinist did not necessarily hold that animals and men were mere automata. But he did hold that organic no less than inorganic phenomena are determined, that is, invariant in the relationship between cause and effect. By contrast the vitalist "entrenched himself behind the word *vitality* and could not be made to understand that saying something was due to vitality amounted to calling it unknown." Only by renouncing the false, merely verbal, accounts of things could true reasons for them be found, and the only way to find them was by experiment:

> If we mean to build up the biological sciences and to study fruitfully the complex phenomena which occur in living beings . . . we must first of all lay down the principles of experimentation, and then apply them to physiology, pathology, and therapeutics.[2]

Bernard, it is clear, had a conception of theoretical biology which differed little in principle from Faraday's conception of theoretical physics. In this he foresaw the future development of his science more truly than any of his contemporaries, and understood the necessary conditions for it.

Despite the inescapable trend of the nineteenth century towards mechanistic forms of thought in biology—though this was not always conscious or explicit—there were important investigations that seemed for a time to favor the vitalist attitude. Chief among them were those of Pasteur, a great experimental biologist who by no means shared Bernard's views; indeed, his life's work was based upon the vindication of a revised, limited vitalism. From his first experiments on optically isomeric crystals he acquired the conviction that such crystals could not be synthesized in the laboratory: optical activity was an organic attribute. Discovering that amyl alcohol (a fermentation product accompanying the normal ethyl alcohol) is optically active, Pasteur set out to prove in his first biochemical investigations (1855–1860) that fermentation is an effect of the life cycle of an organism such as yeast. The chemists, led by Berzelius and Liebig, had at first maintained that yeasts were mere catalysts. Pasteur showed that their hypotheses were untenable.* The same vitalist outlook inspired Pasteur's work in disproof of spontaneous generation; the lesson was always the same: fermentation and growth of microorganisms did not occur unless some living seed were first injected into the pure medium from dusty air or other infected source. The barrier between organic and inorganic was never to be transgressed.

In these researches Pasteur proved that the organic chemists in reacting against *Naturphilosophie* had adopted naive mechanistic hypotheses that failed to work. By contrast, in the field of microbiology his own modified vitalism served as a constructive, fertile inspiration. However, Pasteur had not proved (as he thought) that nothing to be denoted as "living" can arise from nonliving matter, still less, that there is any truly "vital" activity responsible for unique effects. In the long run, Pasteur's work was less significant for its wide implications than for the practical benefits that microbiology brought to medicine, beginning first with Pasteur's own application of his method of immunization to anthrax and rabies, and continuing in the researches of those who followed him like Lister and Koch.

It did not escape Pasteur that microorganisms are the simplest living things accessible to science, nor that his own work was linked with that of the cytologists, as yet chiefly interested in the cells of the metazoa. For, as so often happens in science, the cell theory had developed into a specialty before it had exerted a wide influence on general biology. There is no mention of cells in Darwin's *Origin of Species,* and it was not until the early twentieth century that evolu-

* Just as Pasteur was wrong in his views on isomers, so was Liebig in his view of fermentation. Yet it is now known that fermentation is effected by enzymes which are not life-forces but chemical compounds acting much like catalysts. Meanwhile, to make a clear distinction between "living substance" (e.g., an amino acid) and "chemical compound" has become impossible.

tionary theory and cell theory were united in the new science of genetics. Nor did the experimental physiologists employ the cell concept. And it was only after the publication of Virchow's book *Cellular Pathology* (1858) that it effectively entered medicine. These delays are easy enough to explain. The notion that living matter is almost invariably segmented into many types of small units, called cells, particularly characterized by a special organ, the nucleus, effectively originated in botany with Robert Brown (1773–1858) and Matthais Schleiden (1804–1881). The step of extending it to animals was taken in 1839 by Theodor Schwann. In no sense was the cell theory *discovered:* at a particular moment what happened was that a generalization or scheme of things was declared to be universally true. Yet the inductive justification for this generalization was relatively slight and the interpretation of much of it by the founders of the cell theory turned out, within a few years, to be false. Though the new theory was supported by hitherto impossible observations made with the recently improved achromatic microscopes, its genesis is to be sought elsewhere, in free speculation of the German *Naturphilosophen.** The declarations of the 1830s were manifestoes rather than records of accomplishment. The early cell theory was a brilliant theoretical concept—the first that began the unification of biology—but the elucidation of the working of the cell had still to be made, as had the reinterpretation of all that had previously been known about living things in terms of their structure.

Conspicuously the latter was essential to physiology. Explanation in this science had straddled uneasily across its bases in anatomy and organic chemistry. For despite the pioneering efforts of Malpighi (c. 1670) and the enthusiasm of Bichat, as yet little was known in histology, the study of the tissues of organs and their function. Future histology was destined to be based on the cell, but this had not come about swiftly. The cell itself had to be investigated first, apart from the problem of the cooperation of individual cells in the organism as a whole. Thus the mid and late nineteenth century saw the thorough exploration of the origin of the cell (it was Virchow who coined the phrase "Every cell arises from a cell"); of the parts of the cell, especially the nucleus; of the process of cell multiplication by division; and of the

* At the beginning and for many years nearly all cytological work was done in Germany.

egg and the first stages of embryological growth as special cases of cell multiplication.

This branch of cytology in time provided the experimental background for the modern science of genetics, when the significance of Mendel's theory was at last appreciated. By about 1850 the fertilization of the egg cell was being seriously studied; in the 1870s the essential nature of fertilization and of the early development of the fertilized egg was clear, and by 1890 the movement of the chromosomes had been traced. The significance of the sequence of events thus disclosed as occurring in the early stages of the development of the embryo was not at all certain, however. Obviously the material (whatever it was)that controls the proper development of the new individual from the egg and brings out its inheritance from both parents must be within the nucleus of the egg cell. By 1890 this material had been identified as the chromosomes. So far there was still nothing to explain either reduction–division or the properties of heredity. One influential speculation, that of August Weismann (1834–1914), already declared—in 1885—that the inheritance factor must be a "substance with a definite chemical, and above all, molecular constitution,"[3] but Weismann's concept of "germ-plasm" continuity postulated direct passage of this substance from parent germ cells to those of the next generation. The further development of Weismann's hypothesis failed to account for the fresh observations accumulated after 1885. It was not until 1900 that Mendel's work was revived by H. de Vries and others; then it became apparent that Mendel's experiments agreed exactly with the cytological observations. His "particulate" (i.e., discontinuous) characters corresponded to the particulate nature of the chromosomes, and the pairing of characters postulated by him corresponded to the observed pairing of chromosomes in the cell. As W. S. Sutton (1876–1916) wrote in 1902: "the phenomena of cell division and of heredity . . . have the same essential features, viz., purity of units and independent transmission of the same."[4] Sutton moreover remarked—as was fairly obvious—that the actual agents of inheritance (called *genes* in 1909) must be both more numerous and smaller than the chromosomes with which they are associated.

The experimental work in the United States of E. B. Wilson, Sutton, and especially T. H. Morgan led to a rapid improvement of technique and a much fuller understanding of inheritance, now given a secure place in the cell. Thus, sex determination and the mechanism of sex-linked inheritance were fairly soon unraveled. By 1915 Morgan's school had plotted the position of 50 genes on the chromosomes of *Drosophila*. By 1930 T. S. Painter in Texas had identified genes with bands on the salivary chromosomes of *Drosophila*. Mutation was also studied intensively at the cytological level: by 1925 the first artificial mutations had been formed by X rays; these, H. J. Muller explained as resulting from fracture and resorting of the chromosomes. Meanwhile, exploration of the chemical structure of inheritance, begun as far back as 1869 when F. Miescher (1844–1895) discovered a powerful organic acid (nucleic acid) in the nuclei of cells, was advancing swiftly. In the mid-1930s nucleic acid was localized in Painter's bands; from this and other studies it was concluded that this substance must actually be the chemical material of inheritance at which Weismann had guessed.

Thus genetical problems were now basically biochemical or even biophysical ones. Though it is tempting to trace its antecedents much earlier in "medical chemistry," "animal chemistry" (Liebig), and—much more directly—the "physiological chemistry" of the second half of the nineteenth century, biochemistry proper is a creation of the twentieth. Early nineteenth-century studies had largely been concerned with the assimilation of nourishment by the organism; by 1837 it was clear, for instance, that plants grow by absorbing carbon dioxide from the atmosphere, and that to this process the presence of a green pigment, chlorophyll, is essential. The importance of nitrogen—mostly derived from the soil—became clear only rather later. At the same time the role of nitrogenous substances (proteins) in the nutrition of animals was recognized: it was Liebig who grouped the main constituents of both plant and animal substance into three classes: proteins, carbohydrates, and fats. At this time and later it was invariably supposed that only plants could synthesize such substances, which were broken down by animal metabolism. Bernard's discovery of glycogen in the liver was the first stroke against this hypothesis. Later, many examples of animal synthesis became obvious. In these metabolic studies much depended upon the discovery of reliable tests (such as Fehling's test for the presence of sugar in urine, 1848) * and of special apparatus such as the kymograph and the equipment used for gas analysis. Finally, it should be added that ancient controversies over the seat of metabolism

* This analytic test facilitated Bernard's study of glycogen and the investigation of the then fatal disease diabetes, leading up to the discovery of insulin (1921).

in the animal were ended in the sixties, when it became clear that it was to be found in the cell. In the words of Carl von Voit (1831–1908) in 1881, "the mass and capacity of the cells of the body determine the height of the total metabolism."[5] Yet, even at that time, the *concept* of the cell was still far from limpid, in its relation to the organism as a whole and the totality of its function. To the early theorists the cell was without any doubt the individual, the true center of life, and the organism a collection of such sovereign entities. In mid-century this view was already opposed by T. H. Huxley, and at the close of the century reaction against it brought renewed discussion of the whole concept of cellular structure. Within recent years it has been said that "the most natural unit of life is the living organism" and that the concept of the cell as such a unit "is both unnecessary and unsatisfactory."[6]

Just as the ineffectiveness of biology inevitably produced at the beginning of the nineteenth century a sense of medicine's impotence, so conversely the new power, range, and confidence of biological knowledge at the opening of the twentieth gave strength and promise to medicine. Some of the strength came from improvements in clinical diagnosis arising from the introduction of such new instruments as the stethoscope, ophthalmoscope, cardiograph, and so forth —medicine was very slowly becoming quantitative in its methods—as well as new chemical tests; some of it also from the purer or synthetic drugs for which physicians were indebted to the organic chemists. Chemistry was equally the parent of the two great modern steps in surgery: anesthesia and asepsis. Without these the mechanical and biological refinement of the surgeon's skill that began at a great pace about 1900 would have been quite impossible, and appendicitis (for instance) would have remained a deadly condition. Yet these new strengths of medicine (to which the physicists contributed X rays and radium) hardly penetrated to fundamentals. They made medicine a more sophisticated and successful form of tinkering with the delicate human mechanism, but it was still tinkering.

That it should ever be more than this was still a promise, to which physiologists, biochemists, bacteriologists, and psychologists had pledged themselves. On the one hand there was being built up a thorough knowledge of the normal working of the body, from the organic through the cellular to the biochemical level. Thus disturbances of this normal working could, in principle, be rectified through understanding of the mechanisms involved in their imperfections, rather than through the blundering, empirical methods of traditional medicine. Such was the ideal constantly entertained from the time of Bernard. On the other hand, other groups of scientists attacked the problem of the causes of disease within and without the body, isolating the microorganisms responsible and their vectors, discovering means of protecting population from their ravages. Whereas the old humoral medicine (scarcely obsolete in 1850) had, in all its forms, always tended to discourage the notion that diseases are communicated to sufferers, the new germ theory regarded all noncongenital diseases as transmitted. Thus it was possible to block disease by blocking the route of transmission, or, if this could not be done, the disease could at least be controlled in some cases by suitable immunization procedures, limiting to insignificant proportions the number of human reservoirs of infection. In advanced countries smallpox and diphtheria have been controlled by this latter method and the control of polio is well in sight. The former method of extermination has been most used recently in primitive countries to control malaria, yellow fever, typhus, and other insect-borne diseases.

All this has given medicine, and social technologists, immense power to control not only the external environment in which men live, but also the "internal environment" (as Bernard called it), the complex internal network of relationships and states. The ultimate outcome of the present unplanned, unsystematic progress is obviously obscure. Quite apart from the question of minor, small errors due to ignorance and overenthusiasm (as in the dangerous use of radium in the early days) the human race is entering upon a phase totally different from anything that prevailed in the past. Even its genetic future—and hence the intelligence, character, and physique of the future population of the Earth—are within its control. More slowly, yet surely, men are acquiring knowledge and the potentiality of control in the nervous, intellectual, and psychological realm. All this is good, and has been good, in so far as it has lengthened life and freed it from suffering. That much medicine could do and has done during the last century without changing basically the character of human life on this planet. What is possible, although it can neither be foreseen with certainty nor prevented by any confident foresight, is that by purely biological means, imperceptibly perhaps, this character may be changed without any occurrence of violence or catastrophe. Such a change may prove in some way now impossible to define "better"; it may (as some geneticists argue on statistical grounds) prove obviously worse. These are questions which a doubtful and remote posterity will answer. What is almost certain, however, is that the present easily beneficial effect of scientific medicine on society—making people live longer more pleasantly—cannot continue much longer to operate without vast complications.

Chapter 15

NEW LIGHT ON OLD MATTER

The discoveries of Galvani and Volta placed a new tool in the hands of experimental physicists: the Voltaic pile or electric battery. This new source of electricity immediately stimulated a new series of experimental investigations and demanded fresh theoretical concepts. Parallel with the new science of electricity appeared new theories of light. The early nineteenth-century preoccupation with the unity of nature, so characteristic of German "nature philosophy," suggested the possibility of relating the hitherto mysterious phenomena of light, electricity, and magnetism. The attainment of a "correlation of physical forces" as the mid-nineteenth century physicists called it, was to be the culmination of classical physics; at the same time it gave unsuspected insight into the structure of matter.

The first problem of Voltaic electricity was to determine the origin of the electric current produced within Volta's pile. One view found it in chemical action; Volta himself believed it lay in the mere contact between the two different metals, with the necessary moisture serving to make the contact complete. On the whole, opinion inclined to the former view, strengthened by the knowledge that water had already been synthesized by the detonation of an oxygen–hydrogen mixture by an electric spark. Now the reverse process was tried and water was decomposed by the passage through it of an electric current. The use of Voltaic electricity for chemical decomposition and analysis spread rapidly; it was most profitably used by Humphry Davy (1778–1829), immediately after his appointment to the Royal Institution * in 1801. Davy, decomposing one metallic salt after another, produced a host of new materials, of which the most spectacular were the metals sodium and potassium. Thereby he not only extended the chemist's realm but extended Lavoisier's definition of an element as the last product of *chemical* analysis to include the last product of *electrical* analysis as well. That which withstood both chemical and electrical attack might well be held to be of a simple and fundamental nature.

It is hardly surprising that Michael Faraday (1791–1867), the bookbinder's apprentice who entered upon a scientific career as Davy's personal assistant and amanuensis, should have begun this career with a strong bent towards researches on the borderline between chemistry and physics. Not the least important result of his interest (though by no means the earliest) was his enunciation of the laws of electrochemical decomposition. Careful quantitative study convinced him that the amount of any substance decomposed by electrolysis was proportional to the amount of electricity passed through it, and further, that though the weight of each substance decomposed by a given amount of electricity varied with the substance, this amount was proportional to the *equivalent* weight of the substance in question (p. 261). The concept *quantity of electricity* was by now familiar, since Ohm's Law had been enunciated some years earlier. Ohm (1789–1854), a hard-working teacher of physics in a boys' school at Cologne (already the history of physics was tending to revolve primarily about professional scientists), had been interested in studying the relative conductivity of metals. A series of experiments had taught him that conductivity varied not only with the kind of metal used, but with the cross section of the wire. Ohm sought to devise experiments which would permit him to measure accurately the relative intensities of currents; in the course of this work, adapting to electricity the concepts of fluid flow in common use for the matter of heat (caloric) he defined difference of potential or electromotive force, intensity of current and resistance to

* The Royal Institution was founded in London by Count Rumford in 1799; it was intended as a center for the popular diffusion of applied science. With Davy, Thomas Young, and Faraday it became a combination of research laboratory and a center for the popularization of scientific discoveries in fashionable society.

conduction. The result (in 1826) was his eponymous law, that the electromotive force is equal to the amount of current multiplied by the resistance ($E = IR$). Simultaneously Ampère (1775–1836), professor of physics at the École Polytechnique, the great engineering school at Paris, investigated the influence of currents upon each other, and the way in which direction of flow is influenced by various external forces.

Many factors combined to suggest that the new Voltaic electricity might be found to have close connections with magnetism. It had long been known that iron bars struck by lightning were magnetized. Many late eighteenth-century scientists agreed with the philosopher Kant that there was some underlying connection between all the various forces of nature; as *Naturphilosophie* this became a dominant philosophy in many German universities. *Naturphilosophie* is best known in connection with biology, but it influenced many physical scientists as well. So Oersted (1777–1851), professor of physics at the University of Copenhagen, tried for years under this influence to ascertain whether an electric current had any effect upon a magnet; in 1820 he discovered that if a wire carrying an electric current is placed parallel to a magnetic needle, the needle turns until it lies at right angles to the direction of current flow. Reversing the direction of current flow reverses the direction of the magnetic needle's deviation.

The publication of this experimental proof of the influence of electricity on magnetism attracted immediate attention in the scientific centers of the world. In Paris, Ampère explored the problem further, and declared that he could produce the appearance of magnetism by means of electricity alone. (He did not yet venture to believe that he had produced magnetism from electricity.) In the laboratory of the Royal Institution, Oersted's experiment was repeated by the young Faraday, who also tried the suggestion (made by Wollaston and others) that a magnet might cause a conducting wire to rotate—without, initially, any success, though it was known that a conducting copper wire attracted iron filings in chains perpendicular to the axis of the wire, as if a magnetic field (to use an anachronistic term) existed there. Faraday's delight at his ultimate success, achieved by suspending a conducting wire in mercury so that it could rotate around one pole of a magnet, coupled with the discovery that if the wire were fixed the magnet would rotate, emerges clearly from the pages of his diary. Convinced that Oersted was correct, and that electricity and magnetism were intimately related (though without any trace of Oersted's somewhat mystical faith in the unity of nature) Faraday attempted on several occasions over a number of years to devise an experiment which should permit him to produce electricity from magnetism, once again searching for the converse of a known instance, and much influenced by Ampère's belief that all the known effects of magnetism could be produced by the interactions of electric currents. At last, in 1831, a short, intensive series of investigations led him to discover that it was indeed possible to produce electricity from magnetism, though only intermittently. For example, if a bar magnet was thrust into a wire helix connected to a sensitive galvanometer, there was a momentary deflection of the galvanometer needle as the magnet moved, which ceased as the motion of the magnet ceased. An even more dramatic demonstration was obtained when a copper disk was rotated between the poles of a powerful horseshoe magnet; a current was induced as the disk turned. Faraday announced his results in 1831 in a paper entitled "Magneto-electric Induction," followed by another describing the production of electricity from terrestrial magnetism alone. From this it was only a step to a "magneto-electric machine"—a simple dynamo. Meanwhile an American, Joseph Henry (1799–1878), who had been experimenting on the construction of an electromagnet, had also been attempting to produce electricity by magnetism; he achieved initial success substantially before Faraday, but failed to develop it until after Faraday (quite independently) had published his work.

Faraday turned from his study of the relation between electricity and magnetism to a detailed study of electric induction. He was convinced that the medium must play an important role in induction as it does in conduction, and therefore investigated the electrical properties of both conducting and nonconducting materials, concluding that this difference was not absolute but one of degree only. His discovery that the dielectric (or insulator between two conductors) had an effect upon induction, the charge as it were "spreading into" the dielectric, served in Faraday's mind to dispel the notion of action at a distance, and illuminate that of a field. Later, he also sought for the relations between both electricity and magnetism and light which his view of the unity of nature taught him to expect. An electrostatic field had no effect on the passage of light, but in 1845 Faraday found that a strong magnetic field rotated the plane of polarization of light.* In the same series of experiments he also discovered diamagnetism.

Perhaps Faraday would never have considered the possibility that light might bear some resemblance to electricity and magnetism had it not been for the fact that experimental investigations of light had drastically changed the scientist's view of the nature of this most elusive entity. It had recently been shown that light was not always visible; secondly, it had been related to heat; and thirdly, it had been shown to be a chemical agent. In 1800 the astronomer William Herschel had examined the solar spectrum by means of a thermometer and found that the hottest part lay beyond the red. Whether these "infrared" rays were more properly radiant heat or invisible light was not clear; the latter became more probable after the chemists Ritter and Wollaston had independently discovered the existence of "ultraviolet" rays at the other end of the spectrum. Though these invisible rays were not hot, they showed marked chemical activity, being responsible for the blackening of silver chloride. The connection between infrared rays and rays of visible light was nearly established in the 1840s by Melloni (1798–1854) and more completely thereafter by Tyndall and Samuel Langley (p. 301). All this had the effect of suggesting that visible light was only a portion of the spectrum, and that its nature was not as thoroughly understood as had been thought to be the case.

Had Faraday held the view that light consists of streams of corpuscles (the most common eighteenth-century opinion) it is improbable that he would ever have tried to connect light with electricity or magnetism. But the corpuscular theory of light had been strongly attacked at the beginning of the century, and a veritable revolution in optical theory had occurred by about 1850. Thomas Young (1773–1829) approached the subject as a medical man with an interest in the physiology of vision; thence his interest in natural philosophy (on which he lectured at the Royal Institution in 1801 and 1802) led him to investigate the physical cause of color, particularly the color of thin plates (Newton's rings). He held that any explanation of these based on the assumption that light was composed of a stream of material corpuscles was unsatisfactory, and moreover pointed out that the corpuscular theory failed to explain why intense light

* The parallel effect on the spectrum, for which Faraday also searched, was observed by Zeeman in 1896.

travels at the same speed as weak light. Young saw that if light possessed wave motion then, in a fashion analogous to the beats produced with sounds, two waves could meet in such a way as to "interfere" with one another, and this interference could explain Newton's rings and the colors of diffraction gratings (i.e., of surfaces covered with fine scratchings). As Young described it, "when two undulations, from different origins, coincide either perfectly or very nearly in direction, their joint effect is a combination of the motions belonging to each," just like waves in the sea which, meeting at a channel, either reinforce or cancel one another (Figure 11). In confirmation, Young showed how two sources of light could be combined to give less light than either alone: he made two pinholes very close together in an opaque screen,

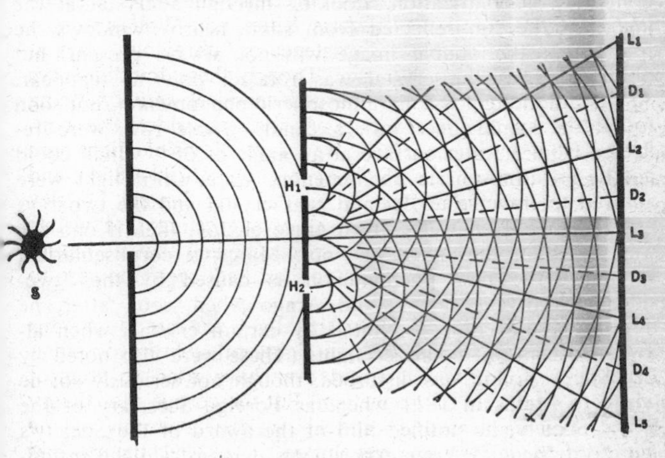

Fig. 11. The scheme of Young's diffraction experiment (not to scale).
Light from the source *S*, passed through a single pinhole, is transmitted through the pair of pinholes H_1, H_2. The "crests" of the waves radiating through H_1 and H_2 are represented by the two sets of arcs of circles. The lines along which the "crests" (and also the "troughs") intersect so that the two waves are in phase are H_1L_1, H_2L_4, etc. The dotted lines show the out-of-phase condition where the "crest" of one wave coincides with the "trough" of another. Where the two sets of waves are in phase, as at L_1, L_2, etc., a light patch is seen; where they are out of phase, as at D_1, D_2, etc., no light is seen.

with a small distant light source, and found that whereas when looking through either pinhole alone he saw a bright continuous light, when he looked through both together he saw the bright and dark parallel bands characteristic of diffraction. It was an exciting discovery but not, it was generally felt, a convincing argument in favor of the wave theory; for the disadvantages emphasized by Newton remained. That Young erred in assuming that diffraction was necessary for interference was not relevant to contemporary opinion.

Nearly ten years after Young's paper of 1801 to the Royal Society, a prize subject set by the French Academy of Science on the mathematical theory of double refraction caused Étienne Malus (1775–1812) to try a series of experiments with an Iceland spar crystal, and thereby to discover the phenomenon of polarization. Looking through a crystal at the image of the sun reflected from some nearby windows, he found that the double image was not always present, but disappeared as the crystal was rotated, only to reappear. At first he thought this an atmospheric phenomenon, but soon found the same effect with a candle whose rays were reflected from a water surface at an angle of 36°, which could not be an atmospheric phenomenon; moreover if light were allowed to traverse an Iceland spar crystal and the two rays allowed to fall on water at an angle of 36°, then if one ray was reflected, the other was not. Malus, a corpuscularian, explained the new phenomenon as caused by the "two-sidedness" of the ray, as did Arago when, soon after, he discovered the colors exhibited by certain crystals when allowed to transmit polarized light. (These were also noted by Brewster.) Young was disturbed, though not seriously so; he wrote to Malus in 1811 when, as Foreign Secretary of the Royal Society, he notified him of the award of the Society's Rumford Medal: "Your experiments demonstrate the *insufficiency* of a theory (that of interferences) which I had adopted, but they do not prove its *falsity*."[1]

Nevertheless Young was somewhat discouraged, more especially since contemporary scientific opinion was flatly opposed to the wave theory. Such leading scientific figures as Herschel in England and Laplace in France wrote on Newton's rings and the double refraction of Iceland spar, respectively, without mentioning Young's theory and upholding stoutly the corpuscular view. Young's interest was revived in 1816 when Arago (1786–1853), who had long been familiar with Young's work, told him of some experiments recently performed by himself and Augustin Fresnel (1788–1827) which indicated that two beams of polarized light did not exhibit the phenomenon of interference under conditions when ordinary light would necessarily do so. Fresnel had been working for two years on the problem of diffraction, much assisted by the patronage of Arago, an established scientist when Fresnel began his researches. Fresnel also discovered interference—apparently independently (since he did not read English)—and demonstrated that it could exist without diffraction by using two plane mirrors at an angle of nearly 180° as his interfering light sources. Fresnel developed the theoretical aspects far more completely than Young, showing that experiment and mathematical law were in excellent agreement. But Fresnel, like Young, could not explain polarization. Arago's report caused Young to reconsider the problem, and within a few months it had occurred to him that the whole matter became clear if one ceased to think of light as being propagated through the aether like sound in air, but as consisting of a wave motion perpendicular to the direction of propagation. As he wrote to Arago,

> I have been reflecting upon the *possibility* of giving an *imperfect* explanation of the affection of light which constitutes polarization, without departing from the genuine doctrine of undulations. It is a principle of this theory that all undulations are simply propagated through homogeneous mediums in concentric spherical surfaces, like the undulations of sound, consisting simply of the direct and retrograde motions of their particles in the direction of the radius, with their concomitant condensations and rarefactions. And yet it is possible to explain in this theory a *transverse vibration*, propagated also in the direction of the radius, and with equal velocity; the motions of the particles bearing a certain constant direction with respect to that radius; and this is polarization.[2]

Gradually Young came to see that transverse vibration was an essential to a satisfactory wave theory of light, even though it was thus impossible to conceive of the aether as an elastic fluid, for it rather resembled an elastic solid. Arago showed Young's letter to Fresnel, who immediately saw that the new theory explained the noninterference of polarized beams. It completely explained "two-sidedness" and rendered invalid the objections of the corpuscularians. But so strange did the notion of transverse vibration appear that, though Fresnel espoused it completely, Arago would not, even when the account of their joint experiments was published in 1819. Fresnel subsequently developed at length the mathematics of the theory of vibration in solid bodies, and applied it in detail to the case of light.

The completeness of the conception and the effectiveness of its development soon conquered some convinced corpuscularians, and prevailing opinion began to look favorably upon the wave theory. As so often happens, a decisive test was only performed when opinion was prepared to accept its conclusion. It had long been realized that the question could be settled by measuring the velocity of light in air and in water; for according to the wave theory light should travel more slowly in an optically denser medium, while according to the corpuscular theory it should travel faster in the denser medium. Wheatstone, in 1834, measured the duration of an electric spark by the use of rapidly rotating mirrors (thereby greatly lengthening its path); he suggested that the same method could be used to measure the velocity of light. Under the influence of Arago, several young French physicists explored the problem: Fizeau measured the velocity of light in air with considerable accuracy (1849), using a toothed wheel combined with a rotating mirror; two years later Foucault succeeded in using Wheatstone's method to measure and compare the velocities in air and water. He found, as he expected, that the velocity in air was greater, and that the relative velocities were proportional to the refractive indices of the media.

All this was very satisfactory, but it left physicists with a most awkward "luminiferous medium" or aether, tractable to mathematical treatment but experimentally indetectable and conceptually impossible. The awkwardness was increased by the lack of any acceptable alternative. The most useful view, and the most influential, was that of Faraday: convinced that light bore some relation to electricity and magnetism, he not only suggested the electromagnetic theory of light which, in Clerk Maxwell's hands, was to yield a set of mathematical equations which mark the culmination of classical physics, but also suggested a theory of matter and of radiation from which the modern field theory ultimately stems.

When Faraday discovered how to produce electricity from magnetism he immediately devised an explanatory model based upon what he regarded as pure experimental evidence. When his copper disk rotated between the poles of a horseshoe magnet, he declared, the disk was "cutting magnetic curves. By magnetic curves I mean lines of magnetic force which could be depicted by iron filings."[3] The fact that the "lines of force" could be made visible showed Faraday, as he believed, that there was experimental evidence for their existence, as there was not for the aether. Faraday's subsequent position is a peculiar one: he rejected all concepts not experimentally verifiable, rejected action at a distance, rejected the atom—yet subscribed to one of the most abstract theories of the structure of matter ever formulated. The atomic theory of Boscovich (p. 270) had long been well known in England: in Faraday's version matter was held to consist of centers of force surrounded by lines of force. The vibration of these lines accounted for electricity, light, gravity, magnetism—all the radiant phenomena, as he called them; that all space is filled with lines of force explains why magnetism or gravity is omnipresent in space. (It is this concept which makes Faraday the originator in some sort of modern field theory.) There was no need for an aether, and all forces and radiations had a common origin in a vibration which behaved in an undulatory matter. The whole notion was highly speculative, but very suggestive.* Clerk Maxwell's first papers on the electromagnetic theory of light were entitled "Physical Lines of Force," soon followed by one on "A Dynamical Theory of the Electromagnetic Field" (1864), in all of which he acknowledged an immense debt to Faraday. Yet though he professed to provide a mere mathematization of Faraday's work, he reintroduced the aether as a necessary medium for the transmission of the vibrations and waves which constitute electromagnetic phenomena. The thesis that light is composed of electromagnetic waves was elaborated in his *Treatise on Electricity and Magnetism* (1873); it was to confirm this theory that Hertz performed a series of experiments in 1888. Hertz worked with "electromagnetic waves" (manifestly containing both electric and magnetic components) derived from either a Leyden jar (condenser) or a spark coil; he found these could be made to exhibit all the phenomena of light, including reflection, refraction, diffraction, interference, and polarization. From this derived Marconi's development of radio.

Meanwhile numerous other aspects of light were being investigated; of these the study of the colors produced by an

* The clearest statement is in a short paper in the form of a letter, "Thoughts on Ray-Vibrations," published in the *Philosophical Magazine* in 1846; it contained the substance of an unpremeditated discussion at the end of a lecture at the Royal Institution.

electric discharge through gases and of fluorescence was to produce the most surprising results. It had long been known that under various conditions a partial vacuum might glow. Faraday, studying the flow of electricity through a partial vacuum in 1836, observed the play of color: he found that as the gas pressure inside the tube decreased the color was concentrated in the area opposite the cathode, suggesting that an emanation from the cathode was the cause. Further progress awaited improvement of laboratory techniques; in the 1850s a Tübingen glass blower, Heinrich Geissler (1814–1879) invented both a mercury pump capable of producing a very high vacuum and what came to be the standard form of tube for studying electric discharges through low-pressure gases—variously known as Geissler, Crookes, or cathode-ray tubes. Perhaps the most important result was Julius Plücker's discovery in 1859 that radiation from the cathode caused the glass wall of the tube to fluoresce, and that this fluorescence could be made to move along the wall of the tube by means of a magnet. This suggested that cathode rays (to use the preferred modern term) were nonmaterial electromagnetic emanations related to light, and negatively charged. Even though Hittorf showed in 1869 that the "rays" could be cut off by obstacles placed in their path, he, like other German physicists, continued to speak of cathode rays as a form of electric discharge. This view was still in force in 1893 when Hertz's assistant Lenard published an account describing how, when he replaced a section of glass in a tube with a square of aluminum foil, the cathode rays escaped through it into the air, where, though they could not travel far, they could still excite fluorescence.

The English physicists meanwhile inclined to the view that cathode rays were in fact particles of matter. The most important and intensive study was that by William Crookes (1832–1919). Trying to make a precise determination of the atomic weight of thallium—for the lines of demarcation between chemistry and physics were still far from rigid—Crookes attempted to weigh hot bodies *in vacuo*. This led him to the discovery of various peculiar phenomena associated with the partial vacuum, and this in turn to the investigation of cathode rays. Crookes produced an enormous variety of cathode-ray tubes, of all conceivable sizes and shapes. His results, announced in 1879, showed that while the observed magnetic deflection of the cathode rays suggested that they were negatively charged, they also were capable of exerting mechanical pressure: for if they fell upon the vanes of a radiometer inside the tube, they not only caused the vanes to rotate, but caused a rise in temperature upon impact. Further, when the tube was almost completely exhausted the vanes no longer rotated, showing that the aether could not be responsible. Crookes showed that the "rays" (which he preferred to term "radiant matter") traveled in straight lines and cast shadows, as well as turned the radiometer. Crookes thought that the residual gas must be in "an ultragaseous or molecular state." He did not attempt to define the precise properties of this "new world where matter may exist in a fourth state, where the corpuscular theory of light may be true, and where light does not always move in straight lines."[4]

The view of Crookes, that cathode rays were material, was espoused by J. J. Thomson (1856–1940) at the Cavendish Laboratory in Cambridge. Thomson used a very high vacuum, and accurately measured the amount of deflection of the cathode rays by known electric and magnetic fields. In each case the amount of deflection must be proportional to the velocity of the particles and their charge–mass ratio (e/m). That these rays or particles were negatively charged had already been shown experimentally. Now Thomson measured their charge–mass ratio, showing that it was identical whatever the gas used. It was found to be about one thousandth of that determined for the hydrogen ion from electrolysis experiments. Thomson called his particles "carriers of electricity"; they were subsequently called "electrons" after a previously proposed "atom of electricity" discussed some years earlier.

Meanwhile Roentgen (1845–1923) had been investigating the fluorescent effect of cathode rays. He shielded his tube with black cardboard, in order to be able to direct a pencil of rays upon a screen covered with the highly fluorescent salt, barium platinocyanide. He found that even when the tube was completely shielded his screen still glowed. Naturally wondering if the source of the fluorescence-producing radiation was really from the cathode-ray tube he interposed his hand, and was astonished to observe the first X-ray picture. An intensive study showed that these mysterious X rays (Roentgen's name for them) could permeate most solid objects, in varying degrees: lead was nearly opaque. Roentgen found that his X rays produced fluorescence in a great many substances, including certain calcium compounds, uranium glass, rock salt, and ordinary glass; he also found that they were responsible for fogging photographic plates left near cathode-ray tubes, a phenomenon long observed. Roentgen, in 1895, was still convinced that cathode rays were "phenomena of the aether"—i.e., a form of light—and believed the same to be true of his rays. He found X rays impervious to the action of magnetism, which showed them to be not cathode rays, but produced by them. He could not reflect or refract his rays with optical equipment; but they did in effect cast shadows, which seemed to indicate some relation to light. Though Roentgen's reasoning was in many respects erroneous, his conclusion was of course correct.

Roentgen's announcement of his discovery created immense popular interest. The medical applications were obvious. But the general public seemed to view the new rays as providing a weapon for evil-minded scientists. There was a fear that there would now be no privacy possible. Berlin newspapers spoke of the new rays as permitting the penetration of brick walls and the end of privacy in the home; a London firm advertised X-ray-proof underwear. The height of fear and absurdity was reached when a New Jersey Assemblyman introduced a bill to prohibit the use of X rays in opera glasses in theaters!

Inevitably, there was keen interest in the new rays in the scientific world as well, and numerous attempts to discover other sources of X rays. Various phosphorescent and fluorescent salts were examined; in particular Henri Becquerel (1852–1908) in 1896 investigated a fluorescent uranium-potassium double salt (the sulfate), to see whether after illumination by sunlight it would emit X rays. He wrapped a photographic plate with enough black paper to prevent penetration by the sun's rays, placed some fluorescent crystals on the paper, and exposed the apparatus to the sun for several hours. Development of the plates showed clearly that the fluorescent crystals had been emitting rays. Further, if a sheet of glass were interposed between the crystals and the paper, the photographic plate still was clouded where the crystals covered it. One day, after the apparatus had been arranged, the sky clouded, and remained cloudy for two days. At the end of this time, as a precaution, Becquerel developed the plate and found that, in spite of the fact that the whole thing had been kept in a dark drawer and no fluorescence was visible, the photographic plate showed a stronger image than usual. A rapid series of experiments convinced him that this was a genuine phenomenon, common to all uranium salts (and to the metal) and that it was unaffected by changes in the environment.*

Two main lines of investigation were now pursued: the chemical, involving the discovery of new radioactive substances (see p. 283), and the physical, concerned with the attempt to discover the nature of the emanations. The principal phenomena associated with radioactivity were the affecting of a photographic plate, the exciting of fluorescence in some substances, the electrification of the surrounding air, and the generation of heat. Of these the first three are also characteristic of X rays; the last therefore appeared the most remarkable. It was soon found that radioactive subtances did indeed produce X rays: there were three main types of radiation (called α, β, and γ rays), and Villard, who first discovered γ rays, identified them with X rays. The first two types of radiation, which Ernest Rutherford (1871–1937) discovered and christened in 1899, were intensively studied for some time before their identity was revealed. Rutherford found that α rays were stopped by aluminum foil and even by air, in the latter of which they traveled always a predictable distance; they were positively charged and heavy. For a variety of reasons—including the fact that helium was always found in uranium crystals—they were first tentatively, and then conclusively, identified as consisting of helium nuclei. Before this it had been found by Rutherford and Soddy that radioactive substances "degenerated" as they gave off α, β, and γ rays, producing elements (some, but not all, new) lower down in the periodic table (pp. 279–280). New methods of detecting and counting individual particles made more investigations possible.

The difficulty was that with each new discovery more theoretical problems presented themselves for solution. Why did radioactive substances grow hot as they decayed? How did radioactive elements differ from ordinary, stable elements? Where did the particles that were α and β rays come from? If, as seemed all but certain, they came from the interior of the atom, what was the structure of the atom, now shown not to be truly atomic, for radioactivity split it up into simpler components? Since the atom appeared to be neutral, whereas cathode rays were composed of negatively charged electrons, clearly there must be some positively charged ma-

* Various uranium salts were in commercial production at this time; one was used to produce a yellow color in ceramic glazes, another was used in photography, and yet another in chemical analysis. The element uranium had been discovered and named in the late eighteenth century.

terial to counterbalance the charge of the electron. J. J. Thomson, influenced by the Boscovich atom, and by various nineteenth-century speculations, suggested a number of models: either electrons were dispersed through a positive soup, or there was a large chunk with electrons circling in rings of eight outside. Rutherford, Geiger, and others bombarded a sheet of gold leaf with α particles. Usually the α particles passed through with little deviation, though occasionally an α particle would be reflected almost straight backwards. This suggested that there must in fact be a small positive nucleus; and that when the positive α particles came close enough they were violently repelled. Rutherford in 1911 postulated an atom which was a miniature solar system, with the positive nucleus the sun, and the negative electrons the planets. He calculated that the (positive) charge of the hydrogen nucleus must be equivalent to the (negative) charge of an electron, and that of a helium equivalent to two electrons. In general, the number of electrons would be equal to about half the atomic weight.

But even as he enunciated this model, Rutherford pointed out its deficiencies. It was impossible in terms of classical physics to imagine that in such a system the electrons could revolve without radiating energy; and if they radiated, the law of conservation of energy demanded that the system must be unstable. Further, this model did not explain the spectral lines observed when a substance was heated very hot. In 1913 Niels Bohr (1885–1962) combined the ideas of Rutherford, in whose laboratory he had been working, with Planck's quantum theory (p. 302) in a new and radical solution. Bohr suggested (like Boscovich) that at certain distances from the small positive nucleus there are null or stationary positions or states, and as long as one or more electrons is in such a position it can revolve around the nucleus without radiating. Bohr postulated that the arrangement of the electrons in their orbits must depend on the position of the atom in the periodic table. He further showed that the spectral lines characteristic of each atom could be explained as the energy radiated by an "excited" electron in moving from one state or orbit to another. This model proved very long-lived and eminently conformable to the demands of both experimental and theoretical discoveries.

The Bohr atom did not touch the problem of the nucleus, which was known to be of minute dimensions in relation to the atom as a whole, to contain virtually all the atom's mass, and to carry a charge equal to that of all the surrounding electrons. There was no reason why the nucleus of each element should not be different, nor to suppose a priori that each nucleus was composed of particles. In experiments of 1919, however, bombarding nitrogen with α particles, Rutherford found other, penetrating particles emitted that were tentatively identified as hydrogen nuclei (or ions, otherwise known from experiments on the ionized gas). He proposed the hypothesis that the α particle joined with the remainder of the disintegrated nitrogen atom to form an isotope of oxygen, so that a transmutation of the elements had been effected. Rutherford further suggested, "if α-particles or similar projectiles of still greater energy were available for experiment, we might expect to break down the nuclear structure of many of the lighter atoms."[5] For this and other purposes particle accelerators were designed during the next few years.

As for the hydrogen nucleus or ion (called a proton by Rutherford in 1920), he was confident that this had originated in the nucleus; indeed, it was difficult to imagine any other source. If this were so—and the hypothesis was soon generally adopted, for magnetic deflection experiments confirmed Rutherford's identification—then the nucleus of an atom of atomic number N must consist of $2N$ protons to give the correct atomic weight; this charge would be double that of the planetary electrons, and so it was necessary to insert N electrons into the nucleus also to render the atom neutral. No evidence of such proton–electron pairs could be found: the problem was resolved only by the discovery of a third fundamental particle, of the mass of a proton but without charge —the neutron. About 1930 it was observed that when beryllium was bombarded with α particles an exceptionally penetrating radiation was produced—more penetrating than γ rays, though otherwise similar. It was next shown that this radiation caused paraffin or other hydrocarbons to emit protons. Further work made the assumption that these rays were γ rays less and less tenable; in 1932 Chadwick at the Cavendish Laboratory investigated the problem very thoroughly and concluded that the "rays" could only consist of streams of uncharged particles having the mass of a proton. Neutrons were found to be even more destructive than α particles. They were also useful in improving the concept of the atomic nucleus, since it could now be assumed that this consisted of a sufficient number of protons to balance the outer electrons, while the remainder of the atomic weight was made up by uncharged neutrons. The neutron was the first of a long line of new particles (the positron was discovered in the same year) that was to emerge from the intensive study of radioactivity in the decades after 1930.

The discovery of new forms of matter stemmed directly from the attempt to explore and correlate the well-known forces and phenomena of classical physics—electricity, magnetism, and, above all, light. But it was in turn to point the way to a new physics. Even before the discovery of radioactivity, classical physics had, here and there, shown itself inadequate to provide a completely successful theory of physical phenomena. The question of the nature of the propagation of light grew more acute as the aether became more elusive. The discovery of radioactivity increased the need for a new theory, demonstrating the existence of totally new kinds of phenomena which required totally new concepts. But though, in time, the new theoretical physicist was to make predictions which the experimental physicist dutifully checked and found exact (as in the case of most of the nuclear particles) it must not be forgotten that from 1850 to 1905, at least, it was the experimental physicist who broke new ground and made the discoveries which necessitated the formulating of new theory.

Chapter 16

THE ATOM CONQUERS CHEMISTRY

After Robert Boyle's attempt to place chemical theory upon a physical foundation there followed a long period during which chemists were chiefly concerned to establish the identity of their science. The problem of chemical combination, as the phlogistonists and the later French chemists saw it, was one of substance rather than structure. Equally the analysis of materials did not signify, towards the end of the eighteenth century, any endeavor to find the ultimate limits of matter; analysis was the resolution of materials into so many parts by weight of the prime substances that Lavoisier called elements. Lavoisier's new theory and the new nomenclature transformed chemistry but they did nothing to render it a more *physical* science, rather than a science dominated by the two laboratory techniques of analysis and synthesis. Few chemists were, as yet, interested in recondite questions. To render the list of elements as complete as possible, to analyze materials into their elementary constituents, and to reconstitute by art the materials that nature provides—to most chemists these seemed ambitions wide enough for a science whose exponents regarded themselves as above all empiricist, pragmatic, and practical.

Yet there were generalizations going beyond these limits, and correlations suggestive of physical hypotheses, that could not be ignored. For instance, within a few years of its proposal by J. L. Proust (1754–1826) in 1799 nearly all chemists had accepted his "law of constant proportions," despite the attack raised by Lavoisier's associate, C. L. Berthollet (1748–1822). Proust's law makes the obvious point that the number of chemical compounds is limited; only certain proportions of elements (by weight) will enter into combination. Analysis reveals the same proportions in any sample of the same material. At the same time the German chemist J. B. Richter (1762–1807) and others reinvestigated affinity —the force of varying strength that binds elements into compounds—affinity itself being obviously connected with the "laws" of composition. The tabulation of affinities in turn prompted the compilation of tables of equivalent weights (the equivalent weights of A and B are those that will combine with the same weight of C): for equivalent weight is simply another way of stating relative affinity. From equivalent weight to atomic weight is a logical—though conceptually difficult—step: because as soon as the phenomena of chemical combination begin to insist upon a periodicity, proceeding in jumps and not continuously, then it becomes reasonable to suspect that matter itself is discontinuous, that is, particulate. Berthollet's hypothesis—that an indefinite amount of sulfur, say, can combine with a given amount of oxygen—would be plausible only if sulfur and oxygen were (as regards combination) continuous in structure.

Few had troubled to pursue atomic speculations in the eighteenth century, though chemists and physicists alike assumed that matter was particulate; among the exceptions were the Croatian mathematician and philosopher Roger Boscovich (1711–1787) and an Irishman, Bryan Higgins. Nevertheless, the point about discontinuity in phenomena and structure was appreciated at last by several persons including William Higgins (1763–1825), a descendant of Bryan. He clearly linked the occurrence of multiple oxides of nitrogen with the possibility of multiple combinations of particles (2:1, 1:1, 1:2, etc.), but did not attain the concept of atomic weight. This was the distinguishing originality of John Dalton (1766–1844), the Manchester philosopher, who for this reason and because he saw how *relative* atomic weights could be computed from the tabulated chemical equivalents, deserves all the credit for first introducing atomism into chemical theory.

Curiously enough, Dalton reached his concept of the chemical atom not directly from studies of combination but through his interest in meteorology. Commencing with Newtonian ideas of atomic structure, he found that these could not explain the mixture of gases, nor their solubilities. He inferred, therefore, that atoms were not identical, but that those of different gases varied in size and so in weight. Up to this point Dalton was constructing a physical theory of gases that could be applied to the atmosphere: now—in the summer of 1803—seeking actually to determine the relative weights of atoms, he turned to chemistry, above all to Proust's law and Richter's table of equivalents. Dalton, of course, was particularly concerned with oxygen and nitrogen, which do enter into a variety of distinct combinations; these he studied, in order to determine the change of proportions by weight. He was not now, however, and never was to be a very accurate analyst, and at first especially he made much use of the work of others in exploring his theory. What was truly original in Dalton was not his experiments nor even his belief in the particulate structure of matter: it was his confidence that the atomic concept could be made theoretically useful if the atoms were distinguished by weight. For the first time, a property of the atom could be derived from experiments. Further, for the first time chemistry truly utilized the generally accepted concept of the *physical* structure of matter.

Knowing the proportions in which elements combine by weight to form different compounds, but not the number of atoms in each least part or molecule of the compound, it is impossible to determine decisively the relative atomic weights.* This simple fact was to bedevil chemistry for two generations. Some security could be gained by appeals to consistency: since atoms of the same elements appear in many different combinations, it is possible to derive relative weights by more than one route, which should prove identical. Dalton further adopted rules of simplicity: a unique combination (or the first of a group) he took to consist of atoms in one-to-one linkage, the second in a two-to-one linkage, and so on. This naturally fitted in with the law of multiple proportions in combination that Dalton also recognized: there can be neither fractional proportions (like 1½ to 1) nor fractional atoms (Figure 12). However, Dalton rejected the parallel generalization of J. L. Gay-Lussac (1769–1854) that gases combine *by volume* in simple multiple proportion, just as he rejected (necessarily, in his system) the notion that equal volumes of different gases contain the same number of atoms. (If

* If water is HO, as Dalton supposed, then the atomic weights are as the combining weights, i.e., H:O = 1:8. But the water molecule being H_2O, the atomic weights (H:O = 1:16)) are in a different ratio from the combining weights.

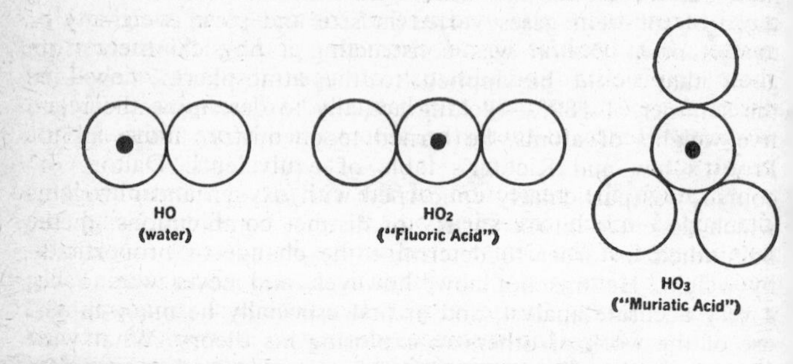

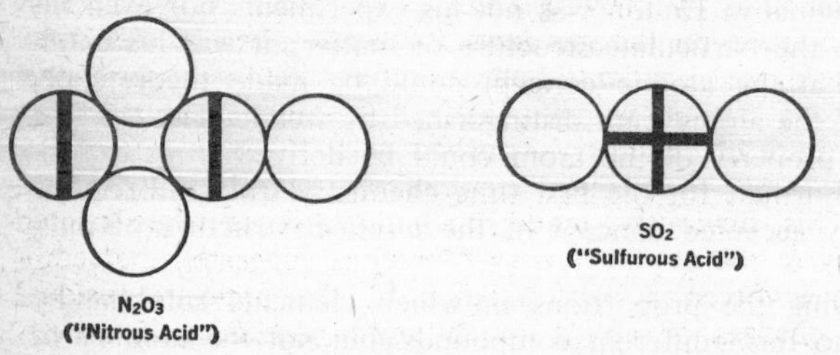

Fig. 12. Some of Dalton's pictures of the combinations of atoms.

Dalton had postulated this, he would have had to set the atomic weight in direct ratio to the density of the gas, which would not have worked out consistently.)

From 1808, when it was described in *A New System of Chemical Philosophy*, Dalton's atomism won the approval of many chemists as offering a satisfactory explanation of empirical truths. Not all followed Dalton rigidly, and perhaps not many allowed the atomic theory to have direct effect on their work. Partly this was because the atomic weights themselves remained uncertain in a peculiarly puzzling way. Aided by such later discoveries as the Dulong-Petit "law" (1819) that atomic weight is given by dividing the specific heat of an element into a certain constant number, and by Mitscherlich's "law" of isomorphism, the great Swedish chemist J. J. Berzelius (1779–1848) had by 1826 determined many atomic weights correctly. Other chemists differed from him, however; complete agreement and consistency in this matter still seemed, at mid-century, unattainable. Consequently many chemists, well into the 1880s, held that while the use of atomic weights was a convenience, the chemical atom was a mere chimera. This attitude seems to have persisted so long not because there was evidence of the complete continuity of matter, but because some chemists, priding themselves on their empiricism and distrust of general ideas, wished to assert the existence only of what could be seen, handled, and weighed. Hence to them the atom was a mere philosopher's construct.

Although the perfection of the system of atomic weights demanded much investigation, it was far from being the sole interest of the near-contemporary founders of nineteenth-century experimental chemistry: J. L. Gay-Lussac in Paris, Humphry Davy at the Royal Institution in London, and Berzelius. Davy, for instance, disproved Lavoisier's oxygen theory of acids by proving chlorine to be an element. Gay-Lussac worked on the gas laws. Berzelius—besides discovering several new elements and the phenomenon of catalysis, working on atomic weights, and contributing immensely to the literature of chemistry—pushed on with the electrochemical hypothesis of chemical combination begun by Volta and Davy. This, obviously, made much more sense when related explicitly to the atomic theory. Berzelius suggested that combination involved a neutralization of the opposite residual electric charges of atoms (each atom possessing, however, both positive and negative charge), and accordingly arranged all the atoms in an electrochemical series. Analysis by electrolysis was conversely explained as a restoration of charge to the separating atoms. Unfortunately, despite the atomic theory and the physical nature of the electrochemical hypothesis, Berzelius suffered from a failure to understand physics. He confused current and potential in a way that led him to reject, as impossible, the equivalence of electrochemical action discovered by Michael Faraday in 1833. From this time his hypothesis depended chiefly on his great authority as a chemist.

There were problems here that penetrated to deeper levels of physical structure than nineteenth-century science was competent to deal with. Meanwhile, experimental chemists—Berzelius and Faraday among them—were also seriously applying themselves to other matters less intellectually profound, perhaps, but highly complex in their detail. These clustered round the composition of organic substances—the complex carbon compounds associated in multifarious ways with living processes or the fossil remains of living things (like coal and oil). It was only towards 1850 that it was generally recognized that organic chemistry was subject to universal chemical laws, not governed by a mysterious "life-force." Organic synthesis was a counterpart of organic analysis: not only because it measured the chemist's skill (or luck!) and promised usefulness, but because the ability to synthesize was a test of the accuracy of the composition found by analysis. The possibility of synthesis, moreover, proved that organic compounds were not unique. The core of the problem, in any case, was to discover the quantity of the elements in an organic substance and—beyond this—to know how a more complicated substance was related, structurally, to less complex ones.

For in organic chemistry the question of structure appeared at last in an inescapable form. It could not be side-stepped. It was no use brewing up correct weights of the elements in the desperate hope that a desired synthesis might occur. Organic molecules had to be built from appropriately shaped bricks. The discovery of isomorphism and isomerism had made this clear by 1830.* In framing theories of structure, chemists had Dalton's grand atomic concept; this, however, had been applied at first only to some simple groups of inorganic compounds. As organic molecules were far larger—those studied about 1830 might contain up to 50 atoms of six different elements—the problem of determining the component atomic weights was increased in proportion. It was rendered more uncertain by the disagreements in principle. And finally Dalton's theory did not at all relate to structural *arrangement*—to the placing of the component atoms in space—a matter which was to become of crucial importance.

Crude empirical formulae (such as C_2H_6O, for alcohol) were derived from reasonably exact analysis combined with the analyst's judgment concerning atomic weights; the transition from these to structural formulae that gave the composition of a unique compound occupied some forty years and the energies of many brilliant chemists. Only at the end of this period (in the 1870s) did it become clear that the nature of a compound could only be said to be known when the position of each of its atoms was plotted. With this problem—not at all envisaged at the beginning—went the secondary one of devising a notation that would do this

* Isomorphism (1819): crystalline shape is independent of the chemical nature of atoms, depending only on their number and disposition. Isomerism (1827): different substances may have identical molecular compositions; e.g., there are twenty-six isomers represented by the molecular formula $C_6H_{14}O$.

readily, as the only way of distinguishing, on paper, between isomers of the same molecular composition. The notion of discovering and recording atomic architecture within the molecule arose, historically, from studies of the patterns which related substances seemed to offer in their composition, and from consequent attempts to classify compounds in order of increasing complexity within a given pattern.

The successful attack on this problem came only much later. For the moment the chemist had to be content with analyses by weight, which he could attempt to translate into statements of atomic composition. First and foremost, every chemist was daily preoccupied with questions of technique—of purification, isolation, analysis, and synthesis. He could only afford to entertain general ideas in so far as he was a master of technique. In dealing with organic materials, because they are complex, the problems of technique were especially serious. Only in the early 1830s, through the work of J. J. Liebig (1803–1873) and his French contemporary J. B. Dumas—the two chemists who dominated chemical research about mid-century—was organic analysis made reasonably exact. It was still cumbersome and tedious. Syntheses were often accidental: the result of breaking a thermometer, or leaving a flame too high. Much in chemistry came from experience, good training in the best laboratories, and an indefinable intuition for how things must be done. Consequently the logical relations between experimental data and theoretical formulae were, for organic compounds, often pretty tenuous even in the 1850s, so that one chemist could see *this* pattern of structural development, another *that*. Sometimes, in turn, the love of pattern overrode the need to observe the facts of (for example) atomic weight.

The transition from "pattern" theories of composition to true structural formulae occupied the middle forty years of the century. Its starting point was the observation of Gay-Lussac that one group of atoms, or "radical," might appear unchanged in each of a series of organic compounds, united each time with a different element. By 1837 Dumas and Liebig, in a joint paper, could proclaim that in organic reactions the radicals played the same part as atoms in inorganic chemistry, identical laws of combination holding for both. This formed one kind of pattern. The discovery that other series of compounds could be formed by exchanging elements *within* a radical allowed another kind of pattern. The recognition of pattern permitted classification, a feature of the organic chemistry of the 1850s in the "theory of types." In this, the elaboration of a simple form into a more complex was traced. So, in the repeated substitution of an ethyl radical for a hydrogen atom, the "ammonia type" appears:

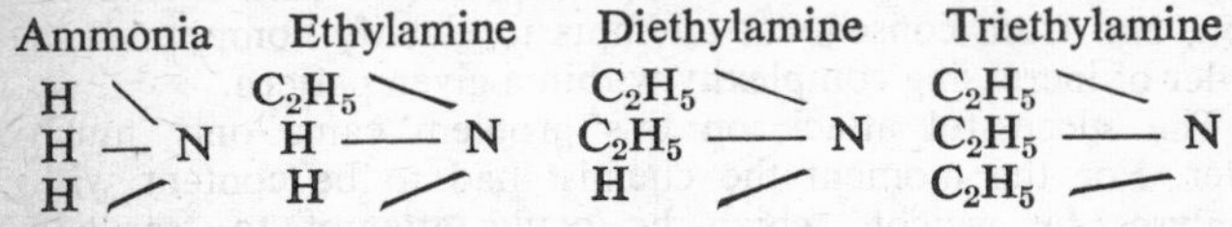

Ammonia	Ethylamine	Diethylamine	Triethylamine
H \	C_2H_5 \	C_2H_5 \	C_2H_5 \
H — N	H —— N	C_2H_5 —— N	C_2H_5 —— N
H /	H /	H /	C_2H_5 /

This representation differs from a structural one in much the same way that Babylonian arithmetical astronomy differed from the Greek geometrical models, but no chemist of the 1850s attempted stereochemistry. What the type theory *did* obviously suggest was prediction from the formulae that such-and-such a new compound should be preparable, and this was found to work well.

The development of the "type theory," which depended upon analysis, empirical knowledge of syntheses, and analogical reasoning, was the work of many chemists. Towards 1880 two especially—the German H. Kolbe (1818–1884) and the Englishman E. Frankland (1825–1899)—had extended it to the treatment of fairly complex substances. The latter, moreover, took a step leading away from a study of patterns towards ideas of the forces between the atoms themselves, which would make possible interpretation of the patterns. He suggested (1852) that "no matter what the characters of the uniting atoms may be, the combining power [afterwards called valence *] of the attracting element is always satisfied by the same number of these atoms."[1] Another German, F. A. Kekule (1829–1896), independently in 1858 recognized the tetravalency of carbon and the possibility, by double-bonding, of constructing a chain of linked carbon atoms, each having three other bonds available for attaching other atoms. One might say here that the striking discovery was that of pictorial symbolism. Once the "attraction" between atoms in a compound was rendered numerical as valence (of one, two, three, etc.) it could be represented by the same number of lines; once these lines could be placed on paper it became feasible to consider their disposition between the atoms, that is, the arrangement of the atoms themselves. How useful such a graphic representation

* In England, *valency*. It was for a time termed "atomicity"—as though to underline the fact that atomism had been half-forgotten.

could be was again plainly demonstrated by Kekule in 1865 when he thought of bending the carbon chain into a ring, as a form suitable for the benzene molecule (C_6H_6), hydrogen being monovalent and carbon tetravalent (Figure 13).

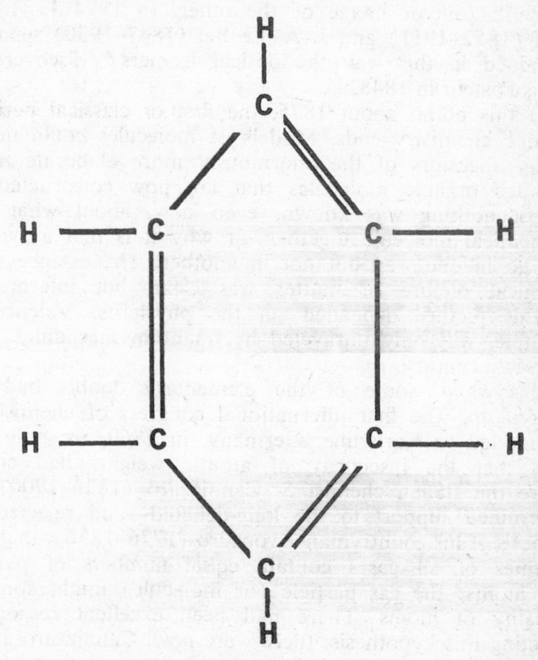

Fig. 13. The benzene ring, consisting of six linked carbon atoms, each with a hydrogen atom attached.

The benzene ring was immediately introduced into the formulae of countless other "aromatic" compounds.

Structural formulae are necessarily two-dimensional; it would be strange if the molecules were so. It was not very difficult to imagine the "atom" as a dense core, surrounded by a three-dimensional frame of valence (replacing the "atmosphere of attraction" of the older atomists). In the case of carbon, for example, the four valences would extend like rods from the central atom to the four corners of a tetrahedron (Figure 14). Similarly in the case of the large molecules: atoms or radicals could be arranged in a three-dimensional picture by combining, with suitable attention to symmetry, the principle of valence and the formulae derived from the type theories. As appears from Figure 14, the data sometimes permit two configurations, one of them being the mirror image of the other. In 1874 J. H. van't Hoff (1852–1911) and J. A. Le Bel (1847–1930) separately explained in this way the optical isomers * discovered by Louis Pasteur in 1848.

At this point, about 1875, the first or classical period in organic chemistry ends. Models of molecules could now be made, ancestors of the enormously more elaborate models of huge organic molecules that are now constructed. But almost nothing was known, even now, about what holds a chemical molecule together, or why it is that a synthesis will go in one way but not in another. The essence of the chemical picture of matter was clear, but interpretation of this picture depended on the physicists. Valence, for example, was only unraveled by quantum mechanics in the twentieth century.

Meanwhile, some of the elementary doubts had been cleared up. The first international congress of chemists was summoned to Karlsruhe, Germany, in 1860, to settle problems that the insecurity of atomic weights had created. There the Italian chemist S. Cannizzaro (1826–1900) gave determined support to the long-debated—and rejected—hypothesis of his countryman Avogadro (1776–1856), that equal volumes of all gases contain equal numbers of particles. *Not* atoms: the gas particle (or molecule) might contain a plurality of atoms. There had been excellent reasons for rejecting this hypothesis; there were now, Cannizzaro argued, better reasons—some of them drawn from physics—for accepting it. Shortly, the majority of chemists agreed and the modern atomic weights came into universal use. Not only did organic formulae become more rational: chemistry now fitted in with the physicist's kinetic theory of gases.

With the atomic weights sorted out, another kind of pattern or classification became more plausible, one that set the elements themselves in order. That this should be possible was also, at the time, a mystery so repulsive that chemists at first derided the whole notion. Yet it proved useful in practice. Ideas of the essential homogeneity of matter as old as Thales were revived by the Englishman William Prout

* Substances, chemically identical, that rotate the plane of polarized light in opposite directions.

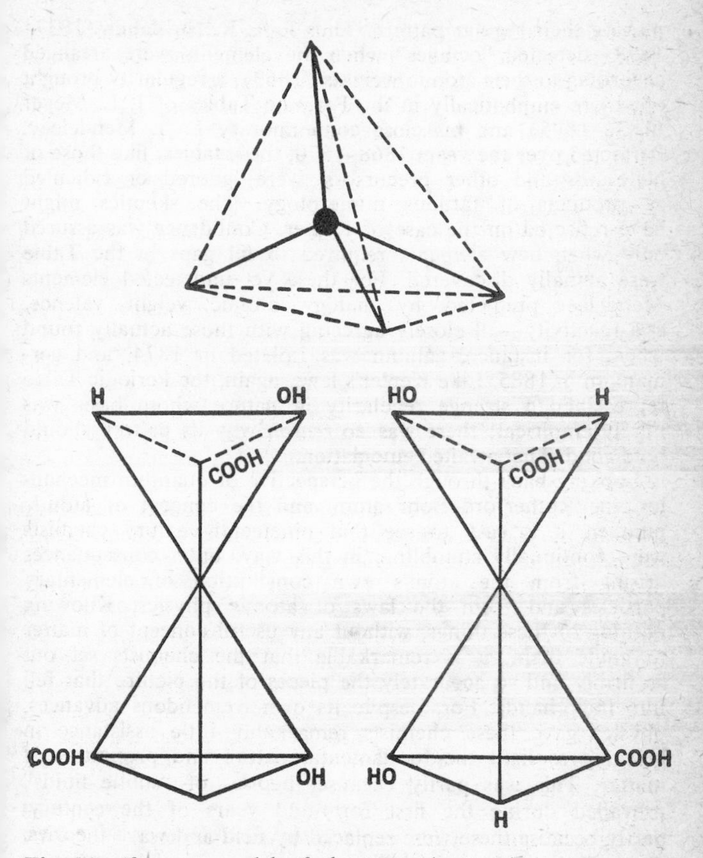

Fig. 14. *Above,* a model of the tetravalent carbon atom as a tetrahedron; *below,* the optical isomeric molecules of *d*-tartaric and *l*-tartaric acids.

(1785–1850), who suggested that all atomic weights should be exact multiples of that of hydrogen, which he took to be the "first matter." Nearly everyone scouted this notion as absurd, pointing to the obvious discrepancies; the existence of isotopes (p. 284) was of course unknown. Although Prout's idea was not pursued, various later chemists drew attention to the recurrence of similar properties among the elements (most obviously, the halogens) and constructed tables displaying their regular pattern. Thus J. A. R. Newlands (1837–1898) detected "octaves" when the elements were arranged according to their atomic weights (1863), a regularity brought out more emphatically in the Periodic Tables of J. L. Meyer (1830–1895) and his close contemporary D. I. Mendeleev. Perfected over the years 1868–1870, these tables, like those of Newlands and other precursors, were ignored or ridiculed as products of fatuous numerology. The skeptics might have reflected on the case of Kepler. Confidence was assured only when new elements required to fill gaps in the Table were actually discovered. For these yet undetected elements Mendeleev predicted, by analogy, atomic weight, valence, and reactivity—all closely agreeing with those actually found when, for instance, gallium was isolated in 1874, and germanium in 1885. Like Kepler's laws, again, the Periodic Table represented a strange regularity in nature whose basis was purely empirical: there was no *reason* why its pattern should hold good. That awaited elucidation.

Looking back through the perspective of quantum mechanics, the Rutherford–Bohr atom, and the concept of atomic number, it is easy to see that nineteenth-century chemists were continually stumbling, in this way, upon consequences arising from the atom's own constitution of elementary particles and from the laws of atomic physics. Knowing nothing of these things, without any useful concept of matter to guide them, it is remarkable that the chemists set out so firmly and so accurately the pieces of the picture that fell into their hands. For, despite its own tremendous advances, physics gave these chemists remarkably little assistance in bringing to light the fundamental nature and properties of matter. This was partly because theories of "subtle fluids" prevailed during the first forty-odd years of the century; partly because these were replaced by field and wave theories. The physicist certainly believed in the particulate structure of matter, but nothing in his theory or in his experiments empowered him to make any definite statements about particles. Hence, before the formulation of the kinetic theory of gases in the sixties, the particulate view of matter (not to say the atomic theory) was essentially a *chemical* view, even though many chemists doubted the atom's reality. If they did so, certainly physicists had no better grounds for asserting it! And it was only in the last quarter of the century, after the enunciation of the Periodic Table, that physics began to move directly into the realm of particles.

Incompatibility with regard to fundamentals did not prevent chemists from deriving advantages from physical discoveries. The most obvious example, in the earlier part of the nineteenth century, is the development of spectroscopy as a tool of chemical analysis. It was to yield many new elements. Despite careful observation of the solar spectrum since Newton's day, it was not until 1802 that W. H. Wollaston (1766–1828) observed some *dark* lines in the solar spectrum which were attentively studied by J. Fraunhöfer (1787–1826) in 1814. Eight years later the *bright* lines of flame spectra were remarked upon by J. F. W. Herschel (1792–1871), but it was not until 1859 that the chemist R. W. Bunsen (1811–1899) and the physicist G. R. Kirchhoff (1824–1887) investigated the matter thoroughly and showed its practical importance. Although they were far from possessing the fundamental theory, Bunsen and Kirchhoff showed that the wavelengths of the lines were fixed for each element (thus serving as a means of identification), and that the bright lines were caused by the emission of light of that frequency, the dark lines by its absorption in the outer layers of the sun. Almost at once Bunsen discovered two new elements, caesium and rubidium.

While inorganic chemistry, the search for new elements, and the determination of accurate atomic weights commanded great attention throughout the nineteenth century, and organic problems were the most exciting in the whole field from about 1830 to 1880, towards the end of the century an increasing number of able chemists turned to the fundamental theory of chemical reaction, whose development required ability in physics, and indeed mathematics. Physical chemistry was chiefly concerned with problems of electrolysis and heat. As for the former, it was early obvious that the passage of current through a solution and the accompanying deposition of its components at the electrodes implied an actual dissociation of the (compound) electrolyte. By 1839 J. F. Daniell (1790–1845) had perceived that the disrupted particles (he called them "ions," after Faraday) could not be the ordinary atoms or radicals of chemistry. He, and others after him, came close to the statement, which was finally made by S. Arrhenius (1859–1927) in 1883, that these ions are charged and remain present in the solution despite the opposition of their charges. This concept was blocked by the seeming impossibility of unneutralized particles existing in a fluid, and of the existence of distinct sodium and chlorine ions (for example) in a solution of salt. Arrhenius, however, had the advantage of intervening studies of dissociation and the movement of ions in solution, which enabled him to present his theory of the immediate ionization of the solution, before the electrodes were applied to it, in an exact and quantitative manner.

As Davy had very acutely guessed, electrolytic dissociation is related to the combining force between elements: other ways of investigating this were developed. One culminated in the law of mass action promulgated jointly by C.M. Guldberg (1836–1902) and P. Waage in 1864. The dependence of the rate of reaction upon the masses of the reagents was, indeed, clearly recognized by Berthollet, who saw that a state of equilibrium is reached the sooner as the concentration of reagents is the weaker. L. Wilhelmy (1812–1864) in 1850 expressed the rate of one reaction (the inversion of cane sugar) formally as

$$\frac{-dM}{dt} = kM$$

that is, the amount of sugar inverted in a given time is proportional to its concentration. This was the first mathematical description of a reaction process. The full statement of the law of mass action is more complex in that it allows for variations in affinity, and for the reverse reaction's tending to restore equilibrium.

The other and far more important investigation of the force—or as we might now say, energy—involved in chemical reactions could begin only after physics had moved from caloric to thermodynamics. This at once suggested, for instance, that Hess's law (1840) * was a case of the conservation of energy. Early attempts to form a theory relating to the heat evolved in chemical reaction, considering this as a measure of the affinity, were somewhat naive; they were improved by Helmholtz in 1882, A. F. Horstman (1843–1929) in 1869, van't Hoff in 1884–1886, and W. Nernst (1864–1941) in 1906. Only then and in a relatively complex mathematical treatment was affinity unambiguously derivable from the heat of reaction. Even then, complex systems required the application of the phase rule formulated by the American theoretical physicist Willard Gibbs (1839–1903). However, although Gibbs announced the phase rule in 1876 and his work aroused the enthusiasm of James Clerk Maxwell in Cambridge, England, the theory remained unknown to chem-

* The same heat is evolved in a reaction whatever route it follows.

ists until about 1890. Thus it was not until about the beginning of the twentieth century that thermodynamical principles were fully extended to chemical reactions.

Nothing so far had shaken the integrity of chemical science as an independent insight into nature and, in so far as nineteenth-century chemistry centered upon the vindication and extension of the atomic concept, this was still in 1900 barely endangered. Chemistry had, indeed, like physics or biology, made important borrowings from other sciences. It had taken up the work of botanists and mathematicians. It had adopted, and specialized for its own purposes, electrical and thermodynamical theories which were becoming increasingly important for understanding the *why* rather than the *how* of chemical combination. By this somewhat derivative means, synthesis—and hence manufacture—was coming to be, about the opening of the new century, a far less empirical, chancy business than in the past. Now, however, chemistry was on the verge of disclosures which would render it no more than a very complicated department of physics. The theory of matter, diverse, incoherent and speculative in the nineteenth century, was to become unified, all-inclusive, and abstrusely mathematical.

All this followed from particle physics, begun in 1874, and was vastly accelerated by the discovery of radioactivity in 1896. At first the isolation and study of radioactive substances was very much in the hands of chemists like the Curies, while interest lay rather in new properties of matter than in a new theory of its structure. From about 1905 the situation changed rapidly under the multiplying discoveries of Rutherford and his co-workers, culminating in the first atomic model of 1911. At this point the intellectual structure of chemistry based on the Daltonian atom collapsed: even the ion—an anomaly since the 1880s—changed its character.

For first, in accord with his discovery of the nucleus, Rutherford's atomic constitution assigned the mass of each atom to the nucleus: this mass was proportional to the atomic weight of each element as determined by the chemists. Here the prime feature of Daltonian atomism was reinterpreted in a subatomic structure. Secondly, and consistently with this, it was observed that the number of "planetary" electrons in Rutherford's model had to be about *half* the atomic weight. In 1913 H. G. J. Moseley (1887–1915) dealt rigorously with this point, defining the *atomic number* of each element (that is, the ordinal number of its true place in the Periodic

Table) as the number of "planetary" electrons in each of its atoms; these increased by steps of one from hydrogen (1 external electron) through helium (2) and lithium (3) to uranium (92). Moseley derived the atomic numbers from measurements of the X-ray spectra of the elements, which exhibited a regular shift through the Periodic Table. Since a slight rearrangement of the order of the elements in the Table to fit their atomic numbers removed a few notable anomalies, it was clear that atomic number was (from the chemical point of view) a more fundamental property than atomic weight, upon which chemists had based the Table.* That is, chemical properties are more directly related to the external electrons than to the nucleus. Finally, Moseley's numbers indicated gaps in the table of known elements: four of the missing elements had been isolated by 1925.

The distinction between the various forms of matter known as elements was now unraveled: it lay in the structure of the atom, in the magnitude of the nucleus and the number of "planetary" electrons required to balance the nuclear charge. At almost the same time the problem of fractional atomic weights was cleared up by F. Soddy (1877–1956). Such weights (e.g., chlorine, at. no. 17, at. wt. 35.46) had led to the rejection of Prout's hypothesis and, after continually puzzling chemists, now consorted oddly with a particulate theory of the atom.

The fact that one radioactive element (ionium), a product of the decay of uranium, was chemically and spectroscopically identical with another (thorium) had been pointed out in 1906. Explaining the formation of the series of radioactive elements by successive disintegrations, Soddy called *isotopes* those elements that possess the same atomic number and chemical properties, but different atomic weights. By extension to nonradioactive elements, chlorine, for example, could be accounted for: any sample contains chlorine of atomic weights 35 and 37 in the proportion of slightly more than three to one. The proof of this hypothesis was given by J. J. Thomson in the same year (1913). F. W. Aston (1877–1945), with the mass spectrograph he devised, soon demonstrated that most if not all elements possess isotopes, and many artificial ones have since been prepared. The true atomic weight of an isotope is always an integer. Among

* There remain a few inversions—elements whose atomic weights do not correspond to the position of their atomic numbers.

the most interesting of the isotopes, "heavy" hydrogen of mass 2 (deuterium) was detected by H. C. Urey in 1932.

Finally, the third dramatic event of 1913—the Bohr development of the Rutherford atom—provided a further theory concerning the disposition of "planetary" electrons that gave an explanation of valence. On the Bohr theory, reactive elements have incomplete outer shells of electrons. Now, the sodium atom (for example) minus one electron forms the stable configuration of neon and becomes electropositive. And the chlorine atom plus one electron becomes an electronegative atom of argon. In combining, the atoms of these two elements transfer an electron and in joining together each attains a stable form. This is electrovalence. In covalence two atoms are linked together by an overlap (so to speak) between their outer shells. If the two shells each become complete by sharing a single pair of electrons (one from each), there is a single bond; if two, a double, and if three pairs, a triple bond. Bohr's theory, applied to chemistry by G. N. Lewis, Irving Langmuir, and others before 1920, not only accounted for the varieties of reactivity classified empirically in the nineteenth century, but also—perhaps more strikingly—for the whole pattern of properties codified in the Periodic Table.

In this way physics yielded up the decisive key to the understanding of chemical combination, and it was now fairly clear how the architecture of atoms was held together. Within less than one man's active lifetime the facts revealed by the development of chemistry during the nineteenth century became fully intelligible. Not that that development had been wholly empirical. To suppose that would be to ignore the strong conceptual element that ran through the search for patterns and regularities which produced both stereochemistry and the Periodic Table. Nevertheless, the theoretical direction of chemistry had been ever weak and uncertain. Its theories had been provisional, *ad hoc,* and short-lived. What was soundly stated was commonly inexplicable; what seemed like an explanation was often false. Some of the greatest chemical achievements (including the Periodic Table) had been rejected at first on intellectual grounds and accepted only because of empirical *force majeure.* As for the atom, the basis of chemical theory, it was no better known than in Dalton's day: the concept had been given no greater precision, for all its increased plausibility.

By the beginning of the 1920s all this was transformed.

True, the atom was no more: split, transmuted, it had dissolved into a mechanics of protons and electrons, to which neutrons and positrons were soon to be added. Chemistry therefore had lost that direct access to the ultimate particles of matter that it had inherited from Dalton. In recompense, however, it was now neatly and firmly tied to physics; not merely to particle physics, but to the quantum mechanics of Planck and the relativity functions of Einstein. A complex synthesis in the organic laboratory was at one end of a line that extended from this majestic mathematical construction. In so far as there was an intellectual order anywhere in the world, chemistry was part of it. The true theoretical development of chemistry begins from this point.

It is worth pointing out that the industrial practice of chemistry took a new turn at the same time, and for somewhat similar reasons. Industrial and laboratory chemistry had been closely related since the seventeenth century at least. Lavoisier was a government official, Joseph Black a consultant to Scottish manufacturers. About the end of the eighteenth century, when the new manufactures of chlorine bleach and artificial soda began, the direction of these enterprises came naturally into the hands of scientifically skilled men. There were few truly empirical inventions in chemical technology, which, by the mid-nineteenth century, was already essential to a large sector of industry. Textiles needed bleaches, dyes, and mordants; the manufacturers of ceramics required pigments and glazes; papermakers, inkmakers, photographers, electroplaters, tanners, metallurgists, and so forth, needed chemicals of all kinds. Few of these commodities were now derived by crude methods from natural sources. Some, like sulfuric acid, demanded for their preparation large quantities of mineral and capacious plants; others were prepared mainly by synthesis.

Upon this already flourishing development, crude and creative of human misery as it was, supervened two others. One was the development of the coal-gas industry, starting about 1810. The ammoniacal liquors, tars, and other by-products of making gas were sources of chemical riches at almost negligible cost. Concurrently, the rapid development of organic chemistry after 1830 offered such potentialities for the synthesis of new materials as the chemist had never glimpsed before. Most industrial chemistry previously had involved either a very simple reaction (like the combustion of the lead-chamber process for making sulfuric acid); or—

as with the treatment of dyestuffs—it was directed to the preparation of natural materials whose chemical nature was unknown; or, as with the Leblanc soda process, it aimed simply to prepare artificially a natural substance. The organic chemists had had experience of all such applications of their science, and at first hoped to do little more than make naturally occurring substances more cheaply than nature. Very rapidly, however, they found their métier in making substances that either do not occur in nature at all or do so in ways that are hardly accessible to men, or useful to them.

As everyone knows, this new type of chemical manufacture succeeded initially with dyestuffs—first the aniline, then the "azo" dyes which are more complex aniline derivatives. The aniline dyestuffs industry springs indirectly from the great Liebig, whose pupil A. W. Hofmann (1818–1892) began to investigate coal tar at Giessen. When Hofmann was brought to London (1845) as professor at the new Royal College of Chemistry, his English students continued the same work. Mansfield, for instance, isolated from it both benzene and toluene, incidentally demonstrating the usefulness of the technique of fractional distillation. Another student, W. H. Perkin (1838–1907) pursued the notion—by no means absurd in the light of contemporary knowledge—that the drug quinine might be synthesized from toluidine. The experiment inevitably failed, but Perkin determined to repeat it on a simpler substance, aniline, from which he again obtained a black mess. However, he was astute enough to observe in this a purple pigment that proved to be fast to light and a good dye. This he proceeded to manufacture, under the name *mauve*.

From this hint, it was not difficult to synthesize other, similar compounds that would serve as dyes; within twenty years a long list of them was available and by the end of the century the production of natural dyestuffs (indigo, madder, saffron, brasil, etc.) had virtually ceased. At the same time, the studies made of both synthetic and natural dyestuffs contributed effectively to the development of organic chemistry. A good part of this work was done in industrial laboratories, especially in Germany, where the structure of pigments was exhaustively investigated.

Next to dyestuffs, the most important new branch of manufacture based on organic chemistry was the pharmaceutical, although the great age of chemotherapy was inaugurated after 1900. Chloroform was discovered by Liebig in 1832 and used as an anesthetic from about 1850; phenol (carbolic acid, discovered in 1834) was first used as an antiseptic in surgery by Joseph Lister in 1865. Such delays in application were at first not unusual, since the medical usefulness of new substances was not systematically searched for; moreover, physicians and surgeons were more conservative in their methods than they are now. Modern chemistry first contributed directly to the pharmacopoeia with the isolation of morphine (1806) and quinine (1820) from the vegetable products opium and cinchona. The advanced study of coal-tar derivatives after 1860 yielded not only dyes and artificial perfumes, such as coumarin, but new and powerful medicaments. Salts of salicylic acid were used as a febrifuge in 1875; aspirin (acetylsalicylic acid) was introduced in 1899. Meanwhile, other antipyretics such as kairine (the first synthetic alkaloid, 1881), antipyrine (1883), and phenacetin (acetophenetidin) had been discovered and found to be of value. Similarly, after the use of cocaine as a local anesthetic (1884), a search for synthetic substitutes was begun which had yielded novocaine and several precursors of it by 1905. The first synthetic hypnotic, chloral hydrate, was discovered in 1868, and barbitone, from which the barbiturates are descended, in 1898. The mass of detail soon becomes overwhelming, but it must be remembered that the relations between chemistry and medicine were still purely empirical. No one knew why these compounds had such-and-such effects, and the effects themselves were discovered only by accident, or by routine testing upon animal subjects. In some ways the benefits to medicine were very great; on the other hand, when so much was unknown, the dangers were considerable too.

Many other products came from the laboratories in the last two or three decades of the nineteenth century: products as disparate as celluloid, explosives, artificial textile fibers, fertilizers, and "vegetable" (hydrogenated) fats. The basis for the *ersatz* element in daily living so prominent in the mid-twentieth century was firmly laid. The organic chemists' promise of riches to come has not been proved false. Since the progress of chemical industry has come to depend to a very significant extent on the discovery of new synthetic processes, while invention of the processes rests upon the possession of an adequate theoretical understanding, it has been to the advantage of industry to promote scientific knowledge by pouring money into both pure and applied research. Even the universities have benefited steadily from this. Correspondingly, science has not merely taught industry how to make new materials; it has also vastly improved the economy and speed of production, chiefly perhaps by the introduction of thermodynamic principles. Finally—as this last point suggests—chemical industry has, in the twentieth century, like chemical science steadily shifted to a firm foundation in physics. Profits may, after all, turn upon electrons and their properties.

Chapter 17

FROM POSITIVISM TO UNCERTAINTY

In the early years of the nineteenth century general idea about the physical structure of the world were vague an incoherent. Apart from Herschel's explorations the univers beyond the solar system was still unknown; it was onl in 1838 that Bessel measured the first stellar parallaxes, an only in 1862 that stellar spectroscopy began. Similarly, Dal ton's atomic theory provided the first slight evidence abou the microstructure of things. Physicists were well aware tha Galileo's injunction to read the book of nature in mathe matics was still imperfectly obeyed, and conscious that th splendid Newton–Laplace architecture stood alone: clear, cer tain, and right. For over a century there was no mathematica physics other than mechanics; the first theoretical structur to join that of mechanics was developed for magnetism an electricity by Poisson and Ampère, between 1810 and 1830 As yet the mental gaps—of which Newton had been full aware—in the physicists' picture of the universe were fille with sketchy, *ad hoc* notions, of which some were ashamed but they knew no other recourse.

Chief among these stopgaps was the concept of subtl fluids: aether, heat (caloric), electricity, magnetism perhaps These were always philosophically suspect, and rightly so They were riveted upon science partly by the old bugbea of action at a distance (for this Newton must bear a fraction of the blame), and partly by naive models associatec with such concepts as "flow," "capacity," and "resistance" that the empirical data required. Lastly, the subtle fluids compensated for the absence of some ideas, like that of radiation, which physics had not yet evolved.

Almost every point concerning the subtle fluids involved a contradiction. They were material yet without mass or volume. They "saved" the simple, impact notion of mechanism yet at the same time destroyed it by making the universe a plenum. The very fact that a plurality of fluids was postulated implied diversity in nature, yet somehow they were supposed to indicate its unity. Even the downfall of the fluids involved a contradiction. For while the first half of the nineteenth century witnessed the disappearance of caloric (matter of heat) it conversely saw the recognition of the wave theory of light, which, more than anything else, made the aether necessary and hence respectable. Radiation (including radiation of heat!) demanded a vehicle for its propagation; the end result—or so it seemed near the end of the century—was not so much that the subtle fluids had been dispensed with, as that they had all been condensed into one. True, there were great conceptual changes, above all in the vindication of conservation principles, but the physical universe was still a plenum.

The seventeenth century believed that the universe was constructed of matter in motion. Heat, for instance, was one manifestation of the rapid agitation of the minute particles of bodies. Consequently, since all phenomena resulted from motion, they could be considered as interchangeable, and further, as Descartes saw, only matter and motion needed to be conserved in the universe to maintain its perfect constancy. Although these ideas were intuitively correct, with a more rigorous development of theoretical mechanics the conservation of motion was shown to be strictly false and for a time the whole idea of *action,* of the dynamism of the universe, was confused under such terms as force, momentum, *vis viva,* and work. It was further associated, in the late eighteenth century and until 1850, with the notion of matter or pseudomatter (the subtle fluids), as though action was always the result of something rolling down a gradient. The real key to thermodynamics, therefore, was the abandonment of this materialization of action, largely under the stimulus of the analysis of the action of the steam engine effected by Sadi Carnot in 1824, and the substitution of the term energy for the term motion to describe the conserved action of the universe, which is never dissipated by the multipl variety of its phenomena.

The seventeenth-century origin of thermodynamics is quit explicit in Helmholtz's classic paper of 1847. "Matter," h declared, "in itself can partake of one change only—a change which has reference to space, that is, motion." In a manner analogous to that of Descartes, Helmholtz argued that if perpetual motion is impossible, the internal events within an isolated system of bodies cannot detract from or add to, the total of forces in that system. In other words, for Helmholtz the conservation of energy is a generalization from the known conservation principles in mechanics that render perpetual motion unthinkable; for if motion/energy like matter, cannot be created from nothing neither can it be annihilated. Had any one ever explicitly denied this? The true novelty of the first law of thermodynamics was its categorical expression of the equivalence and interconverti- bility of all forms of energy, which alone, of course, made the principle of conservation of energy coherent. And the systematic importance of both laws was in their mathematical development and applications.

The mathematical structure of thermodynamics, of which Rudolf Clausius (1822–1888) was the chief architect,* was the second great theoretical edific grappled to that of Newton. Unlike the first—early electromagnetic theory—it was pro- voked by no strikingly novel experimental discoveries, nor was it accepted without a struggle. The "correlation of physi- cal forces" could be discussed calmly; but the principle of the conservation of energy associated with the *equivalence* of "forces" could not gain a hearing when asserted by Mayer in 1843 or by Helmholtz in 1847. J. P. Joule (1818–1889), seeking to demonstrate the equivalence experimentally, was regarded as a hopeless crank. The concept of heat proved, indeed, to be the most obdurate blind spot in the whole history of recent science, of whose existence only certain confusions in the development of mechanics had given a hint. It is difficult to dissociate the idea of heat from simple number scales, and to consider it only as a quantity in equations: as hard as the similar operation for velocity which Gailileo had performed. The seventeenth century had given heat an

* After the pioneer Sadi Carnot (1796–1832), and the physiologist Robert Mayer (1814–1878), the other chief mathematical contributions were those of William Thomson (later Lord Kelvin, 1824–1907), W. J. M. Rankine (1820–1872), and Hermann von Helmholtz (1821–1894).

ir of unreality as a secondary quality: in Tyndall's phrase a mode of motion." But the seventeenth-century kinetic hypohesis had never been rendered mathematical and it had been lrowned by the superficially superior development of the luid or caloric theory in intervening years. Moreover, vhereas "correlation" was a weak word, used in experimental ontexts, "equivalence" was a strong one, used in precise nathematical contexts. To say that heat, mechanical energy, ınd electricity were equivalent and interconvertible was to nake a challenging declaration that seemed, at first, justiied only by a naive faith in the unity of nature. And perhaps omething more was at stake. Even the most extreme of eventeenth-century mechanists, reducing everything to prinary qualities, was left with matter-in-motion: particles buzzng about lay behind all phenomena (if one could but see hem). In thermodynamics it was not so: there was nothing "there" but lumpish matter and protean energy, itself an unpicturable nothingness, a mathematical expression. As Newton had long ago suspected, the mainspring of mechanism was itself nonmechanical. Not only could energy reside, nonmaterially, in space itself, but the transformations of energy were *irreversible,* unlike mechanical ones. As Planck remarked, there was now a "fundamental difference between heat conduction and a purely mechanical process." Compared with the cozy simplicity of the world of subtle fluids, running happily here and there to shed light and bring warmth, the thermodynamic universe was strangely neutral. Metaphoric words like "affinity," "charge," "attraction," had no place in it; with Clausius an anthropomorphic element that had lingered in the underbrush of physics since the time of Aristotle was gone for good.

When thermodynamical reasoning was applied to the least particles of bodies as they were conceived in the mid-nineteenth century, vast accessions of theoretical power and economy were quickly gained. It was no longer necessary to endow atoms with an "atmosphere of heat" and a variety of attractive and repulsive forces. With one exception, the balance sheet of energy transformations was now as complete as Lavoisier had made that of matter transformations, and as Dalton had extended the latter to the chemical infrastructure, so also the mathematical theory of energy was made to work at the molecular level. It was now clear—or seemed to be so in the 1860s—that in this newly perfected accountancy the apparent loss of action in one effect or the apparent gain of it in another were perfectly reconciled with over-all

conservation. Thus the seventeenth-century dream of dedu
ing everything from particles-in-motion was wholly realiz
in principle, and to a large extent in practice. The excepti
was, however, a large and difficult one. Radiations, such
light, obviously involve the dissipation of energy; obvious
this energy derives from its only possible source, the radia
particle. But—in spite of the indubitable triumph of the wa
theory—everything here was obscure and doubtful.

The physicists were the first to generalize about energet
"atoms,"* reviving the long-neglected idea of Daniel Bernou
(1738) that gas pressure could be explained as the hig
velocity impact of the gas particles upon the boundaries th
contained them. Bernoulli had pointed out that (for a giv
quantity of gas) the number of impacts per unit of area
boundary—that is, the pressure—would vary inversely wi
the volume, and that heating the gas (with consequent i
crease of its pressure) could be considered as increasing t
motion of the particles. The investigations of Joule, Cla
sius, and James Clerk Maxwell in the 1850s put these ideas
a rigorous basis. As early as 1847 Joule had derived by
beautifully simple piece of reasoning the average velociti
of gas molecules,† had inferred that changes of phase (sol
to liquid, liquid to gas) were dependent on the velocity, an
had connected this with the latency of heat. His theory, lik
Bernoulli's, explained the gas laws and the absolute scale
temperature proposed by Thomson. In 1851 Joule furth
showed how the specific heat of a gas could be obtained b
thermodynamical reasoning from the velocity of its particle
The number thus obtained—the specific heat per mole—w
only verifiable for what were later known as monatom
gases, the necessary correction for diatomic gases (by allov
ing for the energy taken up in rotation) being proposed b
Clausius in 1857. Clerk Maxwell made the great contributio
of applying statistical methods to the kinetic theory of gases. I
1860 he published his first proof that the velocities of th
particles in a gas are distributed in a Gaussian manner, an
then, deducing the odd conclusion that the viscosity of a g
is independent of its pressure, he offered further confirmatio

* Only a few gases have monatomic molecules. It was this fact whic
caused part of the confusion over atomic weights since it was lor
supposed (wrongly) that the common gases are monatomic.

† 6225 feet per second in the case of hydrogen at 60°F and 30 inch
pressure, or 6055 feet at 32°F. It was later noticed that the veloci
of the particles was about 25 per cent greater than the speed of sour
in the same gas. For the purpose of the calculation Joule assumed th
each molecule had the same velocity.

f the truth of the kinetic theory by confirming his prediction xperimentally. Clerk Maxwell's apparatus is still preserved in ne Cavendish Laboratory at Cambridge, England.

This use of statistical analyses was an innovation of wide-pread significance. Statistical mathematics and the theory of robability were brought into their modern form by Karl riedrich Gauss (1777–1855), who innocently prided himself ipon the total inutility of this branch of mathematics.* Clerk Maxwell had realized that the calculation of the mean ree path for a gas particle turns on the probability of its triking another in a given time. No one could provide firm tatements in kinetic theory by considering the fate of in-lividual particles, yet it was necessary to improve upon he somewhat crude approximations of Joule. A rigorous ki-netic theory must necessarily be probabilistic. Ludwig Boltz-mann (1844–1906), whose work was allied to Clerk Maxwell's in several respects, put the point emphatically:

> A relation between the second law of thermodynamics and the theory of probabilities was first shown when I proved that an analytical proof of that law can be erected only on a foundation which is taken from the theory of probabilities.[1]

The acceptance of the legitimacy of probabilistic arguments marks a great transition in the history of mechanistic ideas. Although it remained true in principle that all the events in a system of bodies were completely calculable, it was now recognized that at the molecular level it was impossible to calculate them. Predictions could be made to determine the future state of a mass of molecules as a whole, but it was impossible (in practice, though not yet in principle) to deter-mine precisely the future state of any one of the molecules. Here was the first "uncertainty relation"—though far less fundamental than Heisenberg's. Here was the first breach in the metaphysical assumption that *mechanistic* meant *deter-minable without limit.*

In time the extraordinarily far-reaching theorems of ther-modynamics were to prove essential to every part of science. In physics the immediate problem was to unite them to electromagnetic theory. Classical mechanics as the center and common origin of all nineteenth-century physical theory did serve to bridge electromagnetism and thermodynamics, and

* However, Gauss himself had used the method of least squares in orbit determination.

to render these three theoretical structures mutually consis- ent; but whereas thermodynamics was related to the classic notion of momentum, electrical theory was more closel connected with that of force. In a sense, they were at op posite ends of the same stick.

That the electrical and magnetic forces were similar t gravitation in observing the inverse square law was demon strated by Coulomb in 1777. In the next generation the grea French mathematicians, who contributed so much to me chanics, took up the theory of electricity, beginning with th distribution of charge on bodies. Coulomb had already con sidered the sphere; Laplace dealt with the ellipsoid; S. D Poisson (1781–1840) developed from 1811 onwards the tru analytic theory of static electricity. His theorems were derive from the assumption that there are two electric fluids, posi tive and negative; more important, however, was his use o the potential function, first developed in the mathematica treatment of gravitational force.* In the 1820s Poisson de veloped a similar two-fluid theory of magnetism, enriching i with the notion of magnetic moment to deal with the dua polarity of the same body. At the same time Ampère formu lated the theory of electrodynamics which taught him to view magnetism as invariably an electrical phenomenon, writin that

> . . . there is no other difference between the two poles o a magnet than their positions with respect to the [mo lecular] currents of which the magnet is composed. . . .

After Faraday's discovery of electromagnetic induction th appropriate mathematical theory was worked out in Ger many, by F. E. Neumann (1798–1895), Wilhelm Webe (1804–1891), and Helmholtz. Thus, by 1850, there was a considerable mathematical structure adapted to disparate ye closely related phenomena, whose unity was asserted no by Ampère's hypothesis alone, nor by Faraday's belief in th correlation of physical forces, but by exact symmetries in mathematical theory.

Now it was obvious from experiments that though gravit and electromagnetic forces alike act on bodies remote fron the center of force, they do so in different ways. Th gravitational field is unaffected by interposed bodies; not s

* Potential expresses the electrical energy in the field surrounding charge: the potential difference between two points in the field—or tw conductors—is the electromotive force.

the electric and magnetic fields, which may be markedly affected. In this respect electricity and magnetism were more like light than like gravity in their propagation; this appeared more clearly later when it was shown that the movement of electricity, like that of light, takes time.* These differences in propagation were not clearly reflected in early electromagnetic theory, which, like Newton's theory of gravity, rested on no physical assumptions about the nature of either electricity or magnetism, or about the manner of their transmission, except that they were fluids. Many of the early theorists were content to suppose that electromagnetic forces might be examples of true "action at a distance" between bodies without the intervention of a physical intermediary, and were thus like gravity. The chief challenge to this view was presented by Faraday, whose hypotheses turned upon the existence of a true field, a continuous medium surrounding the active body through which its activity was transmitted. Thus the electromagnetic fluid, or aether, was rather like the "atmosphere of heat" surrounding atoms at the time when Faraday proposed it. To this extent Faraday's views and his mental imagery of "lines of force" were less sophisticated than a purely mathematical treatment *without* physical assumptions. On the other hand, Faraday's denial of a true action at a distance concept was empirically justified by his own observations on the properties of dielectrics; clearly electricity (and less clearly, magnetism) was transmitted through something whose optical properties—for instance —were thereby modified. By 1852, using such essentially empirical criteria, Faraday had distinguished between *gravity* (not propagated by any means), *radiation* (propagated, but independently of any receiver), and *electricity* (also propagated, but by lines of force dependent on both source and termination). Magnetism he thought akin in its propagation to electricity, though he hesitated to make them equivalent.

Faraday's speculations were not fully coherent, nor were his utterances at different times always consistent. Sometimes he wrote of the lines of force as real—like elastic cords of aether, in Maxwell's phrase—sometimes he wrote as though they constituted only one possible way of representing some state of the aether. If it were merely the case that Faraday's notions induced Clerk Maxwell to devise a fresh, purely mathematical theory of electromagnetism to supplement that

* It had always been assumed—and Laplace had proved—that as gravitation is universal and continuous its action takes no sensible time.

developed first in France, then in Germany, they would be of little importance. The point is, however, that Faraday's arguments and models impelled Maxwell to formulate a *different* mathematical theory, not merely to the extent that he initiated the idea of electromagnetic radiation, but more significantly in developing a true field theory. That is, although Maxwell made his theory independent of the molecular model of the aether from which he first derived his equations, nevertheless the theory was *only* consistent with the idea of a medium through which the electromagnetic waves were transmitted. As Faraday had already pointed out, wave propagation at a definite velocity was incompatible with the idea of action at a distance. Accordingly, Maxwell explained:

> I have preferred to seek an explanation of the facts in another direction, by supposing them to be produced by actions which go on in the surrounding medium as well as in the excited bodies, and endeavouring to explain the action between distant bodies without assuming the existence of forces capable of acting directly at sensible distances.
>
> The theory I propose may therefore be called a theory of the *Electromagnetic Field,* because it has to do with the space in the neighbourhood of the electric or magnetic bodies. . . .[3]

Maxwell presented the complete form of his new theory in 1864. He always expressed his debt to Faraday's conceptions with the utmost generosity, saying that he had at first done no more than express the experimenter's physical hypothesis in a mathematical form. From the equations Maxwell derived it appeared that when the current in a conductor is changed—that is, when the moving charge is accelerated—a wavelike disturbance is radiated into the surrounding field.*

Certain researches of the German, Weber, enabled him to assert that the velocity of the electromagnetic radiation thus predicted was the same as that of light, which in turn suggested that light was an electromagnetic radiation. This was confirmed indirectly by Hertz in 1888. Hertz's researches in turn prompted fresh attention to the mathematical theory of electromagnetic wave transmission and search for

* Hence, the denial of such radiation from the electrons in atomic orbits, subject to accelerations, was a postulated anomaly: the more peculiar in view of the Zeeman effect (1896).

better generators and receivers of the waves. Although the spark remained the only oscillator, communication was improved by tuning the radiation to greater wavelengths with added inductance and capacitance; by using antennae; and by the use of the coherer as a detector. In the last years of the century wireless telegraphy was made practicable by Marconi. Rutherford was one of the early workers on these problems (1897), and nearly anticipated him.

The Maxwell–Hertz discovery of electromagnetic radiation was not merely of enormous interest for effecting the unity of optics and electrodynamics: it induced the bonding of *this* study of radiation with thermodynamics, thus filling the gap left in the theory of energy as it was evolved in the 1850s. Radiation theory in turn became the source of exceptionally important ideas. All this involved the transformation of Faraday's aether from the subject of elastic tensions into the subject of undulations; no less significant than this change, however, is the fact that the aether was still the recipient of energy in either case.* Faraday's implicit definition of a field remained true.

The mathematical theory of the radiation of energy was considerably older than this discovery of a vast spectrum of electromagnetic radiation, which was rendered complete by 1900:

Hertzian (radio), 1888	Infrared, 1800
5000m – 0.5cm	500 – 8×10^{-5}cm
Visible	**Ultraviolet, 1801**
8–3.9×10^{-5}cm	3.9–0.05×10^{-5}cm
Roentgen (X rays), 1895	
c. 0.001×10^{-5}cm	

The experimental work began with Robert Boyle's observation that light colors absorb less heat from the sun than do dark ones—though the connections between radiation and absorption was not definitely stated for two hundred years. The discovery of "dark heat" (infrared radiation) by Herschel

* One of the reasons for rejecting certain early forms of electromagnetic theory after 1850 was that these, appealing to action at a distance, proved irreconcilable with the principle of the conservation of energy.

not only enforced the qualitative separation of heat and light radiations, but suggested an extremely important conceptual device, the "black body" which is the perfect absorber and radiator. However, throughout the first half of the nineteenth century, while the caloric theory prevailed, radiation of heat was interpreted as a simple leaking away of the heat fluid into the surrounding medium or adjacent bodies. It was recognized that heat equilibrium in a physical system entailed equality of exchange between the bodies, and that heat (and cold too!) could be reflected without loss. Thermodynamic principles gave fresh vigor to the study of radiation after a long period of stagnation, for it could now be deduced (from the principle just stated and that of the conservation of energy) that radiation is the exact inverse of absorption. This truth was expressed in several ways: Balfour Stewart (1858) deduced that in comparison with a "black body" a substance will absorb at any given frequency as efficiently as it radiates; Kirchhoff in the next year proved by a "thought experiment" that for any given frequency and with the temperature constant, the ratio between the absorbed and the radiated energy is the same for all bodies whatever; and later still Tyndall showed that *both* the radiation and the absorption of heat by gases are proportional to the number of atoms in the molecule. The equivalence so determined explained the occurrence of bright lines in emission spectra and dark lines in absorption spectra, for a cold substance in the second case absorbs best at precisely those frequencies of light where the same substance, when hot and emitting light, radiates best.

Of this phenomenon, and of the evident periodicity of spectral lines which empirical formulae endeavored to express, there was no satisfactory explanation before the end of the century. Yet it was naturally supposed, from the vibratory character of light, from general mechanical considerations, and such results as Tyndall's, that spectra (and therefore radiation and absorption of energy in general) were caused by motions of the molecules. But to say that light must be akin to heat in being susceptible to a kinetic theory was not to say much.

Newton first obtained from experiments a rule determining the rate at which hot bodies lose heat to their cooler surroundings: the fall in temperature in a given time is proportional to the body's elevation in temperature. This rule was crude, like Newton's measurements. Dulong and Petit offered (1817) a new one,* no less empirical, which in turn was criticized for its restricted range. In 1879, therefore, Josef Stefan (1835–1893) proposed still another relation, of a simpler form: heat loss is proportional to the fourth power of the absolute temperature. Five years later Ludwig Boltzmann provided a theoretical justification of it, based on Maxwell's inference from electromagnetic theory that a radiation (such as heat or light) must press a surface on which it falls. In fact, if the radiation were enclosed within a hollow space having black-body walls, the pressure per unit area must be one-third of the energy per unit volume of the enclosure. Boltzmann (1884) treated the enclosure as though it were a vessel filled with gas, applying this energy–pressure relation to thermodynamics. From this he derived the result that black-body radiation is as the fourth power of the temperature, which was Stefan's law.

At about this time Samuel P. Langley (1834–1906), later Secretary of the Smithsonian Institute and a pioneer of flight, was making measurements of the energy radiated at different frequencies by bodies heated to various temperatures. He used the recently devised optical gratings for refraction and the "bolometer" he invented, which measured temperature by changes in resistance of a platinum wire. His and other work on energy distribution was found wholly consistent with the law of Wilhelm Wien (1893) that the wavelength at which maximum energy is radiated by a hot body is inversely proportional to the body's temperature. This law also was deduced from basic physical principles.

At this point, then, when Clerk Maxwell's electromagnetic theory had been confirmed by Hertz, it seemed also to be firmly bound to thermodynamics. The total structure of physics was fitting neatly together; the number of cross-connections was steadily increasing. The view that radiation is a pumping of energy by an oscillator of some kind ("half is energy of motion, and half is elastic resilience") into a medium; that this energy remains *in* the medium till it is transferred to some receiver, where it produces new physical effects—this had been worked out with mathematical completeness, and justified by thirty years of development in physics. But not completely justified. There remained unsolved problems. One of these was the problem of constructing a *general* theory relating energy, temperature, and wave-

* Heat loss is proportional to $m1.0077^t$, where m is a constant varying with the specimen and t is its temperature.

length for black-body radiation. Wien and others failed to accomplish this. The defect was not particularly serious or interesting as applied to macroscopic bodies within the range of experiment, but for microscopic bodies—the atomic or molecular "oscillators" to which the physical theory of radiation and energy ultimately descended—it was as though it were impossible to relate acceleration, force and mass in mechanics.

The ultimate solution was provided by the German physicist Max Planck (1858–1947), who stands with Bohr, Einstein, and Rutherford as a founder of contemporary physics. His early studies had—unknown to him—been anticipated by the great American physicist, Willard Gibbs; later he worked on the thermodynamic analysis of Arrhenius' theory of electrolytic dissociation. His special interest, however, was in the thermodynamic quantity entropy,* to which little significance was attached at this time, and it was through this unusual approach that he attacked the general theory of radiation. Planck began by investigating a mathematical function which, he found, was proportional to the energy of radiation when the energy was small, but to the square of the energy when it was large. He therefore devised (October, 1900) a new expression containing terms proportional both to the energy directly and to the square of the energy, which proved to represent accurately the known measurements of wavelength, temperature, and energy. Recognizing that this formula had only "formal significance"—like Copernicus' epicycles to Kepler—Planck began on the very day of its discovery to search for its true physical meaning. This, he says, "automatically led me to study the interrelation of entropy and probability" [4]—in other words, to pursue the line of thought inaugurated by Boltzmann. But Boltzmann, although he had insisted on the probabilistic basis of thermodynamics, had had never encountered such strange ideas as Planck was now compelled to develop, despite protracted efforts to reconcile his new theory with the classical concepts of physics. For Planck could not avoid the related conclusions that (1) the energies of molecular-level oscillators—postulated as the sources of radiation from the surfaces of bodies—were not continu-

* Introduced by Clausius, entropy is the measure of the nonavailability of energy in a system; an irreversible change, i.e., one involving the conduction of heat, causes an increase in entropy. Clausius stated the Second Law in the form "The entropy of the universe tends towards a maximum."

ously variable, and that (2) the radiation of energy from these oscillators was discontinuous. The energy of the oscillator could be expressed only as nhc/λ, when n is any integer, c the velocity of light, and λ the wavelength of the radiation. h was a new, universal constant of nature that Planck introduced. hc/λ is the *quantum of energy* associated with the wavelength λ. The existence of radiation implies that the oscillator is sending out energy; clearly this can only happen if the energy of the oscillator falls (instantaneously) from a higher energy level nhc/λ to a lower one, $(n\text{-}1)\ hc/\lambda$, emitting one quantum of energy as it does so. Absorption is simply the reverse phenomenon.

The quantum equation that Planck announced on December 14, 1900, a theoretically derived expression connecting wavelength, energy, and temperature, was easily proved equivalent to the satisfactory empirical formula he had discovered only three months before. But his theory was too revolutionary to gain immediate acceptance. Before the beginning of the twentieth century the emphasis in science was entirely on the continuity of nature, even if this could be justified only by the principle of sufficient reason. Hints that things might be otherwise had been suppressed. Thus the discontinuity necessary to atomism had been mitigated by subtle-fluid theories and aethereal hypotheses, which restored continuity to the structure of things. Equally, biologists turned their eyes from the discontinuity implied by the occurrence of genetic mutations. Even when probability considerations entered physics, they were not supposed to correspond to inherent features of the structure of nature. As Planck wrote in his *Autobiography,* the concept of the quantum

> opened up a new era in natural science. For it heralded the advent of something entirely unprecedented and was destined to remodel basically the physical outlook and thinking of man, which, ever since Leibniz and Newton laid the groundwork for infinitesimal calculus, were founded on the assumption that all causal interactions are continuous.[5]

Though there were recent precedents for Planck's abandonment of continuity, their significance was hardly appreciated. Electricity was once—not long before—considered a continuous, homogeneous fluid; yet the notion that electricity is particulate had been put forward in 1874, again by Helmholtz in 1881, and by the Dutch physicist H. A. Lorentz

(1853–1928) in 1895. These ideas were justified by the actual identification of the electron. Powerful support was, however, given to Planck's theory by Einstein's concept of the photon (1905), which rendered light also discontinuous;* indeed, as a radiation, it must be so at the instant of its generation as an electromagnetic oscillation. Planck had imagined that the discontinuity was smoothed out in the field; Einstein now suggested that the emitted quantum of energy remained intact as a "packet" or photon. Energy was not evenly distributed over the whole field.

The evidence that made this step necessary was all quite recent, starting from some observations of Hertz's in 1888. It was found that metal plates gave off electrons ("cathode rays") not only when they were heated, or subjected to high potential differences, but when ultraviolet light shone upon them; in fact the alkali metals (such as the selenium now used in commercial photoelectric cells) did so under visible light. This in itself was not very strange; the energy of the radiation pried loose an electron. The surprising results appeared in 1902, when Philipp Lenard (1862–1947) discovered by measurements of the electrons' velocities that the velocity of escape was independent of the intensity of the radiation that caused it; in fact, the feeblest intensity produces the effect without any time lag for storage of energy in the metal. Moreover the shorter the wavelength, the greater the velocity of the electrons; while if the wavelength of the radiant light exceeded a certain value for each metal, the effect ceased altogether. This critical wavelength was unaffected by the intensity used. In classical physics these anomalies appeared inexplicable, because again they turned upon discontinuities in nature. Einstein's theory accounted for all of them. A photon's energy hc/λ is inversely proportional to the wavelength; the shorter the wavelength, therefore, the more energy an electron can receive. And if the wavelength is too great, there is not enough energy to release the electron at all.

Einstein's "heuristic point of view" was confirmed by the more exact determinations of the data effected by Robert A. Millikan (1868–1953) in 1916, by the discovery of X ray photons, and still further by the actual collisions between X rays and electrons detected by A. H. Compton (1923).

* The term *photon* was first applied in 1926 to describe "this hypothetical new atom, which is not light but plays an essential part in every process of radiation."

hese could hardly be explained except on the assumption ıat X rays possess particlelike properties. Meanwhile, the uantum theory proved essential to the explanation of spec- al lines, another phenomenon of discontinuity since it is vident that the atom radiates only at some particular fre- uencies, not impartially over the whole band. The French ıathematician and philosopher Henri Poincaré (1854–1912) eclared in the year of his death that only Planck's theory ffered a hope of a solution to this question. In 1913, ross-connecting this area of physics again with the picture f the atom constructed by Rutherford, Niels Bohr—then vorking with Rutherford in Manchester—developed a mathe- natical theory of the atom which depended wholly on quan- um concepts, and so brought together for the first time in hysics the experimental study of the structure of matter, the heory of energy and its radiation, the quantum theory, and he electromagnetic theory of wave propagation.* It is hardly urprising that chemistry also gained much from this gigantic ynthesis.

The confirmation and unfolding of Bohr's theory of the tom was to occupy many future years; much of it still tands unmodified. Within less than a decade, however, the enter of interest in experimental atomic physics, and sub- equently in mathematical theory, was to turn to the nucleus nd its enormous latent energy (pp. 317–318). Somewhat similarly, the early work on X rays led to the investigation of X ray diffraction and its application to crystallography. From this in turn sprang a great increase of knowledge both in solid- state physics and in molecular biology. No science follows a predetermined path of development. While the universal law discovered by Planck in its applications by Einstein and Bohr virtually solved the problems of radiation that had been so nteresting for over half a century to at least a few of the greatest physicists, there was still a subtle difficulty left—one which had caused Planck himself to doubt Einstein's develop- nent of his own concept. For if light behaved in radiation nd absorption as a *particle*, it was unmistakably in its trans-

* It is impossible to summarize Bohr's work here. We might point o the importance of his *correspondence principle* in atomic physics, hat is, the correspondence of deductions from both quantum and lassical physics in their common territory, and his elucidation from he theory (basically, since only pertain "jumps" by the electrons in he radiating atom are possible, only certain radiation energies nhc/λ re possible, therefore only certain lines of wavelength λ are possible) f the spacing of the spectral lines for hydrogen that is actually bserved.

mission and normal optical properties a *wave*. The work
Young and Fresnel was not to be undone, yet it seem
impossible to confer on the same entity two sharply contra
ing properties, or rather, to apply to one and the same s
of phenomena inconsistent theoretical treatments.

One way to remove an anomaly is to change the definiti
of the norm. Louis de Broglie proposed in 1924 an invert
generalization of the photon concept that would associa
with every moving particle a group of waves whose wav
length is dependent upon the momentum of the particle
the same way as with the photon. That is to say, any strea
of fast-moving particles should exhibit, in proper experiment
conditions, the diffraction effect hitherto considere
characteristic of light, X rays, and other forms of wav
motion. In 1924 Davisson and Germer obtained diffractic
effects from a beam of slow electrons; in 1927 J. J. Thomson
son G. P. Thomson verified de Broglie's formula for fa
electrons; and since then it has been confirmed for larg
particles. The duality of particle and wave seems to be re
quired by nature, a duality uncomfortable perhaps but in
escapable. It was most uncomfortable for the mathematic
physicists, who were forced into abstractions far mo
complex and remote than anything known in the classic
physics of the nineteenth century. Werner Heisenberg an
Dirac made even more startling mathematical departure
from tradition; then, adopting de Broglie's suggestion
Schrödinger in 1926 developed in wave mechanics a syster
equivalent to each of theirs, yet closer in its sympathies t
the already distant days of Clerk Maxwell.

At this point two developments of the last hundred year
or so in physical science reached their culmination. The ab
struse mathematics of the 1920s destroyed forever the amb
tion that any profound view of nature could be rendered i
mechanical or even picturable terms. There could be no plac
again for the billiard-ball atom, or for the subtle fluid havin
hydraulic properties, or for elastic tensions in the aether. N
conceptions of classical or Newtonian mechanical science
even, could be appropriate at the microscopic level. In othe
words, the ultimate phenomena in physics—such as electro
diffraction or the Compton effect—could not be accounte
for in language of the type: "This strikes that" or "This pul
that." The only explanation of which they were susceptibl
was that inherent in the logic of mathematical equations
That is to say, the only possible explanation is of the form

f (*x*) follows from *f* (*y*) when *p* is greater than *q*." This does not mean that it is hopeless to give *some* comprehension of particle physics in the older, analogical form, as in the characterization: "An electron possesses both wavelike and particlelike properties." What is implied, apart from the obvious vagueness of the characterization, is the absurdity in words of a proposition whose mathematical origin is far from absurd. We cannot really conceive of a particle that is also a wave, any more than we can imagine a sphere that is also a cube: the two possess contradictory properties. Their definitions are irreconcilable. For that reason, because it clung to the picturable, classical physics failed; conversely, it is impossible that the equations of modern physics should be picturable.

Secondly, the physical theories of the late 1920s completed the destruction of the notion that something ultimate is to be found "there," as the ultimate locus of all phenomena, or the "reality" that mathematical equations describe. Ever since discontinuous, mechanical theories of matter had prevailed over the Aristotelian continuum in the seventeenth century, physical science had had a comfortable metaphysical basis in the confidence (derived from the Greek atomists and reinforced by Newton) that in the last resort scientific analysis would terminate in hard, massy, unchanging particles created by God to endure forever. Even the atomic models of the early twentieth century preserved something of this character in the electron and the nucleus, which Bohr's model only partially removed. Wave mechanics and relativity between them (by rendering the aether superfluous and by identifying matter with energy) radically departed from all previous scientific tradition, denying that there is anything "there" as an ultimate physical reality. Of course the distinction between a particle and empty space is maintained, but it is (again) not one that can be rendered analogically, and indeed it cannot be made completely sharp in any case. In a very strong sense, then, the first third of the twentieth century brought physics back to one of its oldest starting points in the Pythagorean tradition that truth lies in numbers. Complex equations had replaced Plato's simple model of a tetrahedral atom, but Platonic and contemporary physics have in common their deduction of physical systems from mathematical formalism. Just so Max Planck echoed Plato when he spoke at the opening of his *Autobiography* of

> the far from obvious fact that the laws of human reaso ing coincide with the laws governing the sequences c impressions we receive from the world about us, [anc that therefore pure reasoning can enable man to gain a insight into the mechanism of the latter.[6]

Many other contemporary physicists—the late Sir Arthu Eddington most notoriously—have endeavored to express a physical reality in purely numerical relationships, so that (a Eddington put it) given one fundamental equation the theo retical physicist could unfold the whole universe as its entail ment. An ambition far more extravagant than Laplace's. An yet to ask for more than this, to inquire what there is behin the equations, as though these described something capabl of direct apprehension, is to seek to reverse the twentieth century revolution in theoretical physics.

The most curious and still the most commonly misunder stood consequence of this revolution was the Uncertaint Principle of Heisenberg, enunciated in 1927. This asserts th impossibility of determining which electron will next "jump" an orbit in an atom, so radiating energy, or which atom o radium will next decay into an atom of radon. Statistically one can say that the half-life of radium is 1,620 years, ye the inability of physics to predict the half-life of an individua atom is not a matter of statistics, but of principle. Heisen berg's Principle is often presented in the form: we canno know the position and momentum of an ultimate particle a any moment with perfect exactness, because the product o the uncertainties involved in determination of either mo mentum or position is equal to Planck's constant, *h*. (Becaus *h* is very small * the principle is not significant for larg bodies.) Then the difficulty is illustrated by pointing out tha any "signal" by which the particle is to be observed so tha its position and momentum may be known will, as a radia tion, interfere with the state of the particle (as in the Comp ton effect). To argue thus, however, is to make th Uncertainty Principle depend on the question of observability as though it asserted no more than the impossibility of in sinuating experimental devices into a system without disturb ing the system (a truth that is almost trivially obvious ir biological experiments). A better justification for the Uncer

* 6.625×10^{-27} erg/sec.

inty Principle is to say that it simply evolves from the quations of wave mechanics; it is a paradoxical aspect of iat particle/wave duality which is already paradoxical. For ow can a wave have exact position and momentum? One annot give wave properties to a particle without, at the ame time, taking away that sharpness of definition that clasical particles possess.

While it is true that the Principle of Uncertainty, or Ineterminacy as it is sometimes called, formally refutes any pplication of clockwork mechanism to microphysics, it may e doubted whether the wider consequences drawn from it re valid. Heisenberg did not mean to declare that the uni'erse is ruled by chance, nor was he accomplishing the vicory of Free Will over Necessity. The debate of moral, thical, and religious issues is not to be confused with scientic questions about the structure of matter; the absurdity of lemonstrating from a physical theory that man is a spiritual eing is sufficiently obvious. On the other hand, Heisenberg's rinciple did conclude the influence of a philosophic position n science which, if not completely dominant through the ineteenth and early twentieth centuries, was certainly so owerful as to be a major ingredient in every general disussion of the scientific picture of the universe. To some xtent this attitude of mind, positivism, has roots with modrn science itself; to assert the need for empirical proof of cientific statements is to take the first step towards it. Hence rancis Bacon, when he dismissed scientific speculation as so nuch airy word-spinning, spoke like a positivist; so did ialileo when he said science was not concerned with causes but he qualified this by saying, *for the present*); so did Iewton when he argued that he was not proposing theories iat might be debated, but necessary inferences from facts hich could only be upset by proving the facts to be wrongly ated. In the early nineteenth century those scientists who sed all the functional relationships associated with Dalton's tomic theory, yet considered the question of the actual extence of atoms merely metaphysical, were positivists. In all iese instances, however, positivism was partial and unselfonscious; Bacon, Galileo, Newton, and the others often laced themselves in decidedly nonpositivist positions. In fact is point of view was first named and codified by the econtric French philosopher Auguste Comte (1798–1857) in s *Positive Philosophy* (1830–1842). Comte argued that there e three stages of knowledge: the theological, the metaphysical, and the positive. In the first stage things are ex plained by referring them to the doings of Gods; in th second they are explained by appeal to metaphysical princi ples or constructs, like gravity and the conservation o energy; in the third they are not explained at all but merel correlated and compared. To exceed the limits of analytica description, or to postulate any explanatory notion not im mediately verifiable in the phenomena, was in Comte's eye to fail to reach the highest, purest form of science. Thi was true positive knowledge—so termed, because it was ab solutely certain and free from all taint of speculation.

The attractions of this philosophy of science, with it emphasis on exactitude in the investigation of facts and it marked distinction between natural science and othe branches of knowledge, are obvious enough. It is flatterin for a physicist to feel, when he asserts that the specific hea of acetone at 0°C is 0.506, that his knowledge is more ab solute and eternal than that of an historian who records tha the Battle of Trafalgar occurred on October 21, 1805 Comte's philosophy was a straightforward (though spurious answer to such questions as: what is distinctive about th scientific method of inquiry? Are scientific statements true And, why is scientific knowledge cumulative? Consequentl many scientists, besides philosophers, felt that Comte offere a certain guarantee of the value of scientific work, and tha if (as Newton had said!) they did not feign hypotheses bu collected pointer readings, analyzed them and compare them, they would construct something of permanent validity Hence theoretical constructs, anything that could not in fairly direct way be demonstrated and measured, like th atom, or the electron, or Freud's unconscious, was to then a figment; not science, but poetry. Distinguishing, as with razor, between *fact* on the one hand and *fancy* on the other the positivists lost sight of two important provisos: we se only what we know, and what we know is limited by th imperfections of our means of sensation and measurement Even if discovering a truth in nature resembled the uncover ing of buried treasure, it would still be presumptuous to sa that we know anything positively, that is not a formal proposi tion. (Galileo had made this clear long before.) The whol value of the uncovered treasure, in fact, is conditioned b our *prior* familiarity with the worth of gold, gems, Rem brandts, Chinese porcelain, or whatever it may be.

Scientific discovery is not like treasure hunting, however

It is more like posing a puzzle, then solving it; in which case, of course, one expects a pretty close correspondence between the set problem and the answer. If science were *just* pointer readings, any one collection of readings would be as good as another—and as meaningless. It is because science has structure, theoretical interpretation, and explanatory ambitions that the pointer readings can be made, correlated, and fitted into theories. What the universe is *really* like we do not know, and it is meaningless to inquire. We can but form pictures of nature to ourselves, which change radically from time to time and are forever incomplete. Each picture which is generally shared among scientists constitutes the professional view of nature at that time; it is in a sense a mosaic of facts, but the pattern of the pieces in the mosaic that made the picture is not itself composed of facts. It is composed of ideas or theories. And the uncoordinated fact, detached from theory, is as much waste as the piece of colored stone for which there is no place in the mosaic.

Positivism was under attack long before the end of the nineteenth century, some of its opponents going so far as to maintain that scientific knowledge is nothing but the product of our way of looking at nature; hence, could we change our mental habits, we should also change the laws of physics and chemistry. This is not very different from saying that if we knew only one set of optical phenomena we should consider light particulate; if only another set, we should consider it undulatory: which is true and is actually reflected in the history of optics. The twentieth century has not only made it crystal clear that the positive philosophy, in failing to reflect the true character of science as it actually is carried on by scientists, propounded one that is very narrow and boring; it has emphasized the fallacy of any view of science that consists in "uncovering the truth" or "disclosing the secrets of nature," except in a very metaphoric sense. Science actually constructs what it discloses; it is made in men's heads, not read off like an ancient inscription or a divine message. One can therefore legitimately turn round Planck's remark about the correspondence between nature and human reason and ask whether it is not science itself that establishes the correspondence, which is therefore neither accidental nor providential. In any case the construction of science has become, in accordance with the progressive discovery of new phenomena, a matter of such highly sophisticated and intricate abstraction that the ascription of "positive knowledge" to a fabric of mathematical equations would be ridiculous.

This is not to deny that the physicists of 1930 actually knew a vast amount more than the physicists of 1830, but merely to indicate that, partly because they knew more, they were more humble and more aware of the true status of their knowledge.

Chapter 18

THE CENTURY OF PARADOXES

Science has revolutionized the twentieth century; no less surely the twentieth century has revolutionized science. Probably the public esteem for science was highest at the end of the nineteenth century, at a time when it promised peaceful communication between nations and the conquest of disease. At that time there were millions to whom a lecture on paleontology or the stars gave a glimpse of the sublime vistas into space and time enjoyed by the bearded sages of science. It was a moment of self-improvement and idealism when, as we know from some of the chief actors of the period as well as from the novels of H. G. Wells, young men approached their careers in science with a sense of almost religious dedication. The science upon which they embarked was orderly and confident. Everyone believed that in physics at least the major discoveries had all been made, so that further research would be a matter of refinement and consolidation. The "limits of science" in the unknowable were clearly set.

Within the first twenty years of this century complacency and security had vanished. Physics underwent the metamorphoses already described: the ultimate particle evaporated, matter and energy became equivalent, the uncertainty principle was enshrined. The Newtonian universe blew up into the expanding universe of curved space-time. Science which had been (seemingly) majestic, rational, and logical in basically simple ways, became full of imprecisions and paradoxes. Classical science had inspired an awe

Of Newton with his prism and silent face
The marble index of a mind for ever
Voyaging through strange seas of Thought, alone.[1]

Modern science (as it is somewhat quaintly called, as though it were "contemporary" furniture) has evoked a different poetic view of travel:

There was a young lady named Bright,
Whose speed was far faster than light;
She set out one day
In a relative way
And returned home the previous night.[2]

And other conceptions less lighthearted. The fantastic dreams that had bubbled since the 1870s suddenly hardened into the reality of human flight, submarines, the hell of warfare conducted with poisonous gases and high explosives, radio, the splitting of the atom, the artificial induction of genetic mutations, and governmental organization of society to a hitherto unprecedented extent. Moreover, while these revolutionary developments were taking place in pure and applied science, the stable social order that had endured (essentially) since the downfall of Napoleon in 1815 was itself overthrown, first by the world war of 1914–1918 and the revolutions in Russia and central Europe, then by the economic and industrial collapse of 1929. Typically, in this century of paradoxes, the era of greatest potential production in the history of the human race was inaugurated by a tragic failure of the means of distribution, which inhibited production.

While the less fortunate peoples who have only in this century won independence and the future promise of prosperity look to science and scientific technology for the comfort and plenty that the West enjoys, and while Marxists still speak of science as the signpost on the road to their future Utopia, the people of the Western nations who have already in some measure gained a Utopia accept the changes of the twentieth century with nonchalant indifference. To them modern medicine is, like television, a commonplace; they take for granted wonders of which no medieval necromancer would have dared to boast, and receive with numb distress prophecies of destruction that would have stirred earlier generations to mad religious frenzy. Perhaps in compensation they tolerate almost without murmur social and political outrages that would have convulsed Victorian, or any earlier, society. Throughout all this, and against the frequent protests of the scientists themselves—who have in general understood little more of the current in which they were borne than the rest of humanity—science has permanently modified its character. It has ceased to be an inquiry into the secrets of nature's laws and processes. It has ceased to be a private or academic pursuit. And many ordinary people have learned to fear it.

There is a contrast of images close to the heart of the tragedy of our time: for to close this book with easy appeal to the riches and ease which future science can bring would be a mockery. (Yet it is true that unimagined resources of goods and leisure lie before us, if we can but grasp them.) One is that of the professional scientist who proclaims his devotion to the cause of truth and knowledge, come what may, and his utter lack of interest in and responsibility for the results of society's use of his discoveries. "I give you the power to make certain choices," he says, "and the choice you have made is yours." The other is the popular image of the Advanced Technician. Probably not one per cent of the population realizes or cares for the distinction between pure science and advanced technology: cosmologists, rocket experts, television engineers, monkey puzzlers, virus extractors, and microbiologists—they are all scientists. What is *this* scientist doing? He is getting to the Moon, making nuclear bombs, putting additives in gasoline, preventing bread from growing stale, and inventing contraceptive pills. Almost the only science that the citizen knows in the mid-twentieth century is concerned with the power to change things; it is concerned not with knowledge but with altering our lives in one way or another, perhaps by ending them. Almost the only scientist that the citizen hears of is not a sage but a man who is an employee of the government or some industrial corporation. The scientist has become *their* servant, not humanity's.

Of course it is impossible to go backwards, futile to long for Jeffersonian democracy or Merrie England. The complexity of twentieth-century science which renders it totally incomprehensible to the vast majority of men and women—not in its intricate detail, but in its very structure and objects—is inevitable and unavoidable. It happens that the world we live in is more complicated than that which classical physics constructed. To explore all the theoretical ramifications built into the model of a large protein molecule, of which the colored balls on wires are (in themselves) no more than conventional symbols, just is intellectually a great deal more intricate than exploring what is represented in a model of the solar system. And of an atom one cannot even begin to construct a model. Correspondingly, as the process of scientific discovery has become more complex—so that nowadays various types of biologists employ thermodynamical concepts, matrix mathematics, elaborate statistical theory, quantum mechanics, X ray crystallography, and so on—the machinery of science has grown exponentially larger, more powerful, and more expensive. In the early 1920s the chief tools of atomic physics were a scintillation screen and a microscope; the first particle accelerators were built just after 1930. Now international cooperation is required to construct the most powerful machines. Of the world's largest telescope there is but one example, at Mount Palomar in California; and even of the 100-inch size, which is half as big, there are only two working. At a much lower level no university or industrial research laboratory can be called well-equipped unless it possesses mass-spectrographic, chromatographic, and other different types of chemical analysis, radiation apparatus of different kinds, ultracentrifuges, and so forth. Of all this the equivalent, sixty years ago, was an ordinary bench microscope, a spectroscope (both made of brass) and assorted simple forms of glassware. Although modern mechanical resources enable research to proceed with vastly enhanced rapidity, many scientists doubt whether the number of really important new ideas has increased and many believe that really fundamental discoveries can only come from physicists who, lacking great machines to work with, turn once more to improvisation and new principles. To these it must seem only proper that in 1962 the Nobel prize-winning physicist Rudolf L. Mössbauer could write of his experiments that: "A day spent in the Heidelberg toy shops contributed materially to the acceleration of the work."[3]

As it is inevitable that the frontier of science becomes ever more remote from common experience, so it is also unavoidable that the social repercussions of scientific and technological developments will have more serious social repercussions. Even when science acts for the best (in preserving life or increasing leisure) this is true. The catastrophic displacements that accompanied the so-called "Industrial Revolution" (c. 1750–1850) afford a grim warning to our own age, in which change is incomparably more swift and profound. Stress of a different, psychological, kind appeared in European society somewhat later, when first Darwin and then Freud challenged on scientific grounds some of the most cherished and personal of human beliefs. Clearly, whatever the future scientific society may be like, its birth is bound to be painful, for it can resemble closely neither our contemporary society nor any romanticized version of humanity in the past.

Again, the military overtones of so much scientific effort in the mid-twentieth century are inevitable products of our history. Not only Western civilization but every other on earth has turned all its resources—intellectual, spiritual, and technological—to warlike purposes, often in the frank pursuit of national aggrandizement. There is every precedent for this use of our contemporary scientific capacity—though less for the moral camouflage under which it is disguised—and there are countless instances since the seventeenth century (and even earlier) of men's trying to devote scientific skills to military ends. The difference between our age and previous ones is chiefly that science was of relatively little use in warfare until the twentieth century, partly because warfare itself was so crude and so much less a total preoccupation of any state. Victorian armsmakers would surely, if they could, have made more use of chemists and metallurgists than they did. Unfortunately, the ability of scientists to show how terribly complete and revolting acts of destruction can be accomplished has come at a moment when not a few people believe that the stakes at issue in world politics today are so high that *any* exercise of power is justified. Hence, in the words of Winston Churchill (1955) "by a process of sublime irony a state [is reached] in this story where safety will be the sturdy child of terror, and survival the twin brother of annihilation."[4] This is Churchill's paradox, "the worse things get, the better." We hope.

Who could have foreseen this state of affairs? It is perhaps the cruelest of all the century's paradoxes that it was a great humanitarian and lover of peace, Albert Einstein, who not only provided the theoretical expectation that vast amounts of energy could be released from the nucleus of the atom, but who signed the letter to President Franklin D. Roosevelt which led to the organization of the Manhattan Project and the explosion of two atomic bombs over Japan in 1945. Einstein was never an atomic physicist; the fundamental equation, $E = mc^2$, predicting that a small amount of mass is the equivalent of a very large amount of energy, was a consequence of Einstein's Special Theory of Relativity (1905), six years before Rutherford presented his picture of the atom. Einstein's equation predicted an evolution of energy roughly a million times more efficient than that obtained by ordinary physicochemical means, and in the late 1930s it became clear that in certain nuclear reactions such a conversion of mass into energy must occur. The rest was a matter of enormously difficult scientific and technical development: now it is known that matter can be converted into heat and radiation by both fission and fusion (or nuclear and thermonuclear) processes, but the latter—the H-bomb—has not yet been tamed for peaceful exploitation.

Einstein's Special Theory of 1905, in which he extended to optical phenomena the concept of relativity restricted to mechanics in classical physics, while maintaining the constancy of the velocity of light under all circumstances—from which two principles it follows that no material body can move as fast as light—was itself the product of certain philosophic views on the character of scientific propositions as well as of the contemporary relation between theory and experimental evidence. The fact that Einstein had liberated himself from the naive appraisal of classical physics—"this describes the world as it really is"—was very significant. In particular Einstein was convinced of the truth of Ernst Mach's contentions that the Newtonian absolute dimensions were by no means necessary to the structure of physics, nor experimentally warranted. Although he was not a positivist, Einstein, like Planck, allowed a higher permissivity in scientific theory than did classical physics, that is to say he admitted into theory (as did Planck) premises and conclusions that seemed absurd by common-sense or classical physical standards, so long as the resultant theory would fulfill the purposes for which it was framed, and entailed no scientific consequences that were demonstrably false. The obvious example is Einstein's abandonment of the Euclidean parallel postulate in the General Theory of Relativity. Hardly any one had doubted until then that the geometry of physical space is Euclidean; Einstein showed the theoretical benefits of treating space as Riemannian, or non-Euclidean, and his predictions on this basis were later verified by observation. Another way of expressing this would be to say that Einstein, in this respect differing completely from older physicists like Kelvin, was prepared to entertain theoretical notions of which no mechanical model could be constructed. And first he abandoned the aether.

As already remarked, the wave theory of light committed physics to the aether, which was at first considered (mathematically) as virtually a solid and then as a liquid with some properties of a solid. Later in the century Faraday's experiments and Maxwell's electromagnetic theory made the aether still less avoidable and still more troublesome. The aether was postulated as fixed in absolute space, so that the molecules of the Earth, for instance, swept through it with a velocity compounded of the Sun's proper motion, the Earth's orbital and diurnal revolution, and so on. Indeed, for the undulatory theory of light to work the aether *had* to be fixed; this was the presupposition (to take one example) upon which Fresnel had explained the "aberration" of light discovered by the English astronomer Bradley in 1729. Fresnel supposed that the observer moved relatively to the aether through which the light waves from the star were traveling. In 1887 the American physicists Albert A. Michelson and Edward W. Morley, attempting an experimental test of Fresnel's hypothesis, established the fact that no motion of the apparatus through the aether could be detected. The accepted explanation of "aberration" was refuted.

A rescue operation was essential. It proceeded through a totally fresh consideration of dimension theory, of measurement, and particularly of the notion of simultaneity. Einstein's Special Theory depends upon a new definition of the word *simultaneous*. The first step was a highly paradoxical hypothesis propounded by the Irish physicist G. F. FitzGerald in 1892; he observed that if the dimensions of the apparatus in the Michelson–Morley experiment were reduced by the factor $\sqrt{1 - v^2/c^2}$ (where v is the velocity of the apparatus with respect to the aether and c that of light) the results of the experiment would actually be consistent with motion through the aether. Or, to put it the other way round, absolute motion through the aether is unobservable precisely because measuring rods shorten in the direction of their translation (though, obviously, this shortening is inappreciable at normal speeds and unobservable at any speed). Precisely the same expression was introduced by H. A. Lorentz of Leiden as distinguishing the comparison of coordinates in his system from the comparison appropriate to the Newtonian system: this modified *Lorentz transformation* was incorporated by Einstein in the Special Theory.

Meanwhile, phenomena hardly less odd were becoming known in a different quarter of experimental physics. The particulate nature of the electron was established by determining the ratio (e/m) between its charge and mass for which the measurements of J. J. Thomson were decisive. However, it appeared that as the velocity of the electron increased, this radio diminished; Thomson, investigating the general case of a charged sphere in motion, recognized that the magnetic energy of the accompanying field could be interpreted as an increase in the mass of the sphere. Henri Poincaré (1900) conjectured that *all* energy possesses inertia—which is, of course, classically an attribute of mass.* Finally, the Special Theory of Relativity gave the relation of mass to velocity as

$$m = \frac{m_0}{\sqrt{1 - \frac{v^2}{c^2}}}$$

m being the mass of the particle moving at velocity v, and m_0 its normal or rest-mass. Here again the "paradoxical" factor that distinguishes the new equation from that of classical physics (in which $m = m_0$ invariably) is the same as that in the FitzGerald contraction and the Lorentz transformation. Further, it is possible to show that the increase in mass $(m - m_0)$, as it is proportional to the rest-mass and the square of the velocity, has the characteristic of kinetic energy such that $E = (m - m_0)c^2$. In its rigorous and fully general form this equivalence enables the twin conservation principles of classical physics (mass and energy) to be replaced, as Einstein pointed out, by a single principle of conservation.

No one in 1905 thought of the conversion of mass into energy as an experimental operation. Only the subsequent experimental and theoretical advances in atomic physics made this possible, and indeed almost thirty-five years passed before nuclear fission was announced on January 15, 1939.

So far we have only traversed one-half of the Einsteinian paradox. The other half begins with the General Theory of Relativity that Einstein proposed in 1915 as an extension of the Special Theory. Its basis was the identification of gravity with inertia. It had of course long been recognized that gravity (regarded in classical physics as an ultimately explicable phenomenon) and inertia (a law of nature requiring no explanation) where alike in many ways, and chiefly in the identity of gravitational mass with inertial mass—an experi-

* Poincaré exercised a great formative influence on twentieth-century physics, and if this book were to extend into even a brief history of the theory of relativity it would be just to pay fuller tribute to his work.

mental identity in classical physics. Since Newton's time it had been clear that exact experiments supported this identity; for example, the period of a pendulum does not vary with the weight or density of its bob; but there was no theoretical reason why this should be so. Furthermore, to render this seemingly strange conflation of gravity and inertia possible, Einstein formulated a new concept of space–time which is, again, relativistic in the sense that the departures from classical physics are significant only when velocities approaching those of light are involved. Otherwise the field equations yield equations of motion identical with those of Newton. Despite the paradoxes which at first inspired many scientists to reject Einstein's work, the General Theory was (in the words of Willem de Sitter)

> a purely *physical* theory, invented to explain empirical physical facts, especially the identity of gravitational and inertial mass, to coordinate and harmonise different chapters of physical theory, and to simplify the enunciation of the fundamental laws. There is nothing metaphysical about its origin.[5]

Like all significant scientific theories, the General Theory solved one major puzzle—in this case the anomalous motion of the perihelion of Mercury, inexplicable in classical terms —and predicted another observable effect (the bending of light by large masses) which was confirmed in 1919. Although many scientists found the abandonment of Euclid incomprehensible and absurd, for those who could penetrate it there was no doubt of the cash value of General Relativity, which in a few years became the basis of new cosmologies (such as those of Einstein himself, and de Sitter) or the target of opposed cosmologies such as those of E. A. Milne and more recently Bondi, Gold, and Hoyle.

Thus from about 1920 onwards General Relativity has furnished a new and immensely important part of the intellectual equipment to be deployed by anyone considering a comprehensive view of the universe. At the same time important material was also pouring in from the new high-altitude observatories using very large reflecting telescopes and other sensitive apparatus.* Besides giving information about the chemical composition of the sun and stars, spec-

* After Herschel, Lord Rosse (1800–1867) took the next step in size with a 72-inch speculum-metal reflector with which spiral nebulae were first discovered (1845). More reflective silver-on-glass reflectors were made from 1880 onwards. The first great observatories of modern times were the Lick (1876), Yerkes (1892), and Mount Wilson, where the 100-inch began work in 1918. Photography of nebulae with large reflectors began about 1890.

troscopy permitted a distinction between galactic and extra-galactic nebulae. Since Sir William Huggins had identified the gaseous type of nebula by means of the spectroscope in 1864 it had been generally agreed that all such objects belonged (like the star clusters) to our own galaxy. From about 1910 evidence was gathered to indicate that the spiral nebulae (and some others such as the Magellanic Clouds in the southern sky) contained discrete stars at very great distances. In the 1920s the study of Cepheid variables in these nebulae by Edwin P. Hubble (1889–1953) made it quite certain that they are true independent galaxies far remote from our own. The Large Magellanic Cloud is distant about 75,000 light-years, and the spiral galaxy in Andromeda about ten times as far again. In comparison, the diameter of "our" galaxy is about 100,000 light-years, its thickness (if it has the spiral form) being very much less.

Photographic studies are determining the distribution of matter in the cosmic spaces; Einstein's General Theory predicts what the character of this space must be. (At high velocities even time assumes well-known paradoxical properties.) Meanwhile, spectroscopic studies of the sun, stars, and nebulae have made it possible to classify stars into groups whose respective evolutions are fairly well understood. An observation, confirmed in the late 1920s, that was not predicted by the basic cosmological theory of Einstein, was the "red shift" in the spectra of distant galaxies, found to grow more marked in simple proportion to the greater distance. If this is indeed (as is generally admitted) an instance of the Doppler effect, then the only possible interpretation is that the universe is expanding so that the galaxies are receding from one another at such a rate that the distance between them doubles every 1300 million years. From this it seems to follow either that at some one point in time (the creation?) the universe began to expand, or that as the volume of the universe has increased so has the quantity of matter in it. This last interpretation, involving the continuous creation of matter (of course at a rate which is imperceptibly slow for volumes of experimental scale), has not been generally accepted by physicists and cosmologists, who have preferred, rather, to maintain the principle of the conservation of mass–energy.* The Einsteinian cosmos, though unbounded

* This preference is less obviously rational than it sounds, for the choice is *either* to maintain the general conservation principle *or* to admit that the universe is not uniform (on the large scale) in either space or time. The latter is a view of the universe that theology ha

(which means that a light signal in any direction from any point would never reach a limit to the universe) is also not infinite (because the light signal travels along curved lines); hence the quantity of matter in the universe also is finite, although its average density in space is (probably) decreased by expansion. Some very general considerations led Sir Arthur Eddington (1882–1944) to compute this total mass of the universe as about 10^{22} times the mass of the sun.

Although in some of his later writings Eddington seemed to play the role of a twentieth-century Kepler, endeavoring to deduce the structure of the universe from considerations of pure number, he was also a hard-headed astrophysicist who contributed greatly to the understanding of stellar composition and radiation. It was long obvious that the sun and stars must derive their radiated energy from their own matter (or they would have cooled long ago) and that this could not be done by any ordinary chemical combustion. The late nineteenth century supposed that the energy came from gravitational attraction, heating the stars by the compression of their contraction. This is still considered an important effect in astrophysics, but it cannot constitute the chief source of the radiant energy. By about 1920, quantum theory and atomic physics had allied to suggest how stellar matter, at its exalted temperatures of millions of degrees, must be in a state very different from any ordinarily encountered in science. At such high temperatures Wien's distribution law requires that the peak radiation be of very high frequency, which has the effect of ionizing the hot stellar matter so that the atoms are largely deprived of their electrons. (This high-frequency radiation is mostly reabsorbed, the main energy from the stellar surface being at lower frequencies.) The remaining atomic nuclei are compressed to incredible densities.

To account for the enormous radiation of energy over millions of years, however, it is necessary to employ Einstein's equation and admit that stellar matter is actually converted into energy. No nuclear reactions making this possible were known in 1920. Eddington suggested that electrons and protons might combine and annihilate each other, but this is impossible; so is the process of nuclear fission, since

always taught and science has resolutely rejected, until recent years. As Hoyle has written, "It is true that we must not accept a theory on the basis of an emotional preference, but it is not an emotional preference to attempt to establish a theory that would place us in a position to obtain a complete understanding of the Universe. The stakes are high, and win or lose, are worth playing for."[6]

the heavy elements are lightly represented in the stars, if at all. The stellar matter consists for the most part of hydrogen and helium, the lightest elements. If four protons (hydrogen nuclei) combine to form one α particle (helium nucleus) plus two positrons, there is a net loss of mass which reappears as energy; moreover, the two positrons combine with two electrons to give further energy in the form of γ rays. Thus the stellar energy can be derived from nuclear fusion, the building up of elements, and indeed it seems likely that the whole series of elements heavier than hydrogen has been so formed within the stars, hydrogen being the primitive form of matter. Hans Bethe, who has contributed so much to nuclear astrophysics, has shown how helium can be formed from hydrogen in stars very much hotter than the sun by a nuclear synthesis through which a carbon nucleus becomes by successive additions a nucleus of nitrogen and oxygen, which, after the last fusion of a proton, divides to yield the original carbon and a nucleus of helium.

Such nuclear reactions can take place only at very high temperatures. Hence when man attempts nuclear fusion he has to use such a very high temperature as is provided by explosive nuclear fission (this, of course, is the H-bomb), or some other means such as the rapid passage through a gas of an enormous electric current. Astrophysics and military technology have already joined hands—a union inconceivable only a generation ago.

The most dramatic and enthralling of scientific adventures, to which man's best endeavors have been devoted, has ended in the threat that the human race may be extinguished. Whatever the future may be—whether or not some future age will view nuclear technology as the most useful and pacific of arts—it will always be true that our age is (and we hope posterity may say *was*) on the edge of disaster. For what this age has done posterity too will pay the price in genetic damage; that no one disputes, though there is argument over its relative significance. Yet, until 1945, what a splendid, untrammeled adventure it was! With an extraordinarily powerful alliance of experiment and theory, both of seemingly endless resourcefulness, physicists were penetrating the deepest secrets of the nature of matter and thereby, almost incidentally, seizing the key to countless other mysteries. It was one of those moments—deceptive of course—when the final capitulation of Nature seemed almost in sight. Everyone knew that the nucleus contained (so to speak) dynamite, but no one yet thought of this dynamite as a destructive explosive.

For various reasons, including the military uses of nuclear energy and the barrier of secrecy and suspicion that surrounds the whole field of experimental high-energy physics, it may well be that this branch of science, having so emphatically marked the twentieth century, has now passed the climax of its intellectual reward. There are signs that the most exciting movements of the second half of this century will take place not in the physical sciences at all, but in biology. Perhaps in this historical way, if in no other, the recent Nobel prize-winning studies of molecular architecture in living things are comparable to the atomic models of the first decade of the century. With the structure of the enormous molecules that make up living things once known, it will be possible in the future to determine the pathways along which they and the still larger organic units, such as cells, are actually formed and how they interact one with another. The early nineteenth-century vital force is turning out to be, after all, but a very complex development of thermodynamics. Thus the bridge between the organization of living things and the physical sciences will become complete.

Although actually creative of vast social problems (through increase of population, for example), and potentially destructive when directed to war, biological science has had a far more positive impact upon mankind than has physical science. More than half the world's population is alive today not because physical scientists and technologists have found work for them to do, but because biological scientists have provided the means to prevent their dying. There is consequently some, if no very rosy, hope that a tremendous surge forward in biological knowledge will hold fewer explosively harmful possibilities than recent physics has done. (No one, however, could have predicted 1945 in 1911.) Indeed, it is clear that biological technology—which is in some respects and in many parts of the world relatively backward as compared with physical technology—can provide the solution to problems that are of even greater human interest and importance than those of energy production, automation, or space travel. In population dynamics, in medicine and public health, in the genetic enhancement of the race, and in food supply, there is urgent need for discussion and action, which is not the same thing as tyrannical control. Some measures are already within our power; others depend on the development of biological research; but it is surely fairly certain that the present chaotic situation whereby the rich peoples get richer and the poor poorer cannot continue forever. We have reached the point where the fantastically productive evolution of the physical sciences upon the Newtonian foundation, transformed in turn by the double theoretical and technological revolution of the twentieth century, offers the advanced peoples a material Utopia of a kind. (Their distress as measured by resort to psychiatry, high suicide rates, and so forth is another matter.) It seems very unlikely that a sophisticated technology offers anything like the same ready promise to underdeveloped countries. Nuclear power is not of tremendous value to a country that gets much of its energy from burning dung, that lacks capital, transmission networks, electric motors, factories, and a trained labor force. It is doubtful whether a very numerous and poor peasant society can rapidly industrialize itself without encountering enormous social, economic, and psychological disasters. For example, Britain, Europe, and the United States industrialized themselves on the basis of a relatively ample food supply in a situation where the availability of food could actually improve rapidly although the number of food producers was declining and the number of consumers increasing. Yet there was starvation. In Asia it is hard to see where the improved food supply to match industrial development is to come from, unless biological technology is deployed along with, and parallel to, physical technology. That is to say, the kind of "scientific revolution" that will best work for social betterment in Asia may well prove to be of a different kind from that which has produced the contemporary societies of Europe and North America.

Perhaps the vital research is going forward at this moment. Who can say? It is quite unlikely to be a part of any investigation whose future effect is foreseen. For not the least of the paradoxes of the twentieth century is its unevenness in foresight. Physicists can draw a curve plotting the course of the sun's radiant energy over the next five thousand million years, during which time it will expand to engulf the earth and then cool to a cinder. Biologists can predict statistically the future distribution of genes. Some scientific predictions are indifferent to the actions of men; others may be altered by men's acts. But of our own future we can foresee very little. A few generations may see Utopia on this planet. They may see standing room only, or the exhaustion of energy supplies; no one can be certain, and most predictions about the future of the race have been falsified by time. On the whole, only a few cosmologists, physicists, and biologists care much about the more distant future. Yet perhaps that is the highest manifestation of the intellectual eminence that science has brought to men.

NOTES

CHAPTER 3

1. Sir Thomas Heath, *Greek Astronomy* (London, 1932), pp. 100-101.

CHAPTER 5

1. Geoffrey Chaucer, *Canterbury Tales,* translated into modern English by Nevill Coghill (Harmondsworth, Penguin, 1951), p. 496.
2. C. D. O'Malley, *Michael Servetus* (Philadelphia, 1953), p. 60

CHAPTER 6

1. Henry Osborn Taylor, *The Medieval Mind* (Cambridge, Mass.), II, p. 379.
2. A. C. Crombie, *Robert Grosseteste and the Origins of Experimental Science, 1100-1700* (Oxford, 1953), p. 69.
3. Lynn Thorndike, *History of Magic and Experimental Science* (New York, 1923), II, pp. 654-65.
4. Geoffrey Chaucer, *Canterbury Tales,* translated into mod-

ern English by Nevill Coghill (Harmondsworth, Penguin, 1951), pp. 479-480.

5. Ben Johnson, *The Alchemist,* Act II, Scene iii.

CHAPTER 7

1. Aristotle, *History of Animals* (London, Bohn, 1862), VI p. 142.
2. Linnaeus, *Systema Natura,* quoted in E. Nordenskiold, *History of Biology* (New York, 1946), p. 210.
3. Pliny, *Natural History,* VIII, 1, Loeb Library, III, pp. 3-4.

CHAPTER 8

1. Arturo Castiglioni, *History of Medicine* (New York, 1947), p. 127.
2. Galen, *On Anatomical Procedures,* Charles Singer, tr. (London, 1956), p. 2.
3. *Ibid.*
4. Realdus Columbus, *De re anatomica libri XV* (Frankfurt, 1593), pp. 326-327. (Translation is by the author).
5. D'Arcy lower, *William Harvey* (London, 1897), p. 62.
6. Paracelsus *Selected writings,* J. Jacobi, tr. (London, 1951), p. 167.

CHAPTER 9

1. A. C. Crombie, *Robert Grosseteste and the Origins of Experimental Science* (Oxford, 1953), pp. 299-300.
2. John F. Dobson and Selig Brodetsky, *Nicholas Copernicus De Revolutionibus, Preface and Book I,* Occasional Notes of the Royal Astronomical Society, No. 10 (London, 1947), Book I, Section 9.

3. Stillman Drake (ed.), *Discoveries and Opinions of Galileo* (New York, 1957), p. 238.
4. Christiaan Huygens, *Treatise on Light,* S. P. Thompson, tr. (Chicago, 1945), p. 3.
5. Robert Boyle, *Works* (London, 1772), I, p. 356.
6. *Ibid.,* IV, pp. 68-69.
7. Isaac Newton, *Optiks* (London, 1931), p. 400.
8. Isaac Newton, *Principia,* Motte-Cajori, eds. (Berkeley, 1946), p. xviii.
9. I. Bernard Cohen (ed.), *Isaac Newton's Papers and Letters* (Cambridge, Mass., 1958), pp. 302-303.

CHAPTER 10

1. Isaac Newton, *Optiks* (London, 1931), p. 404.
2. I. Bernard Cohen (ed.), *Isaac Newton's Letters and Papers* (Cambridge, Mass., 1958), p. 48.
3. I. Bernard Cohen (ed.), *Benjamin Franklin's Experiments* (Cambridge, Mass., 1941), p. 213.

CHAPTER 11

1. Arthur O. Lovejoy, *The Great Chain of Being* (Cambridge, 1948), pp. 190-191.
2. Antoni van Leeuwenhoek, *Alle de Brieven* (Amsterdam, 1939), II, pp. 163-164. (Translation modernized by author.)
3. H. G. Alexander (ed.), *The Leibniz–Clarke Correspondence* (Manchester, 1956), p. 12.

CHAPTER 12

1. J. M. Robertson, *The Philosophical Works of Francis Bacon* (London, 1905).

CHAPTER 13

1. Thomas Burnet, *The Sacred Theory of the Earth,* 5th ed. (London, 1722), pp. 1-2.
2. J. M. Drachmann, *Studies in the Literature of Natural Science* (New York, 1930), pp. 85-86.
3. Asa Gray, *Darwiniana* (New York, 1876), pp. 91-92.
1. M. F. X. Bichat, *Recherches physiologiques surela vie etla mort* (Paris, 1844), p. 56. (Translation is by the author.)
2. Claude Bernard, *Introduction to the Study of Experimental Medicine* (New York, 1949), II, pp. 184-185.
3. Arthur Hughes, *History of Cytology* (New York, 1959), p. 81.
4. *Ibid.*, p. 87.
5. A. Castiglioni, *History of Medicine* (New York, 1947), p. 790.
6. Arthur Hughes, *op. cit.*, p. 143 (quoting Sir James Gray).

CHAPTER 15

1. Thomas Young, *Miscellaneous Works* (London, 1855), I, p. 157.
2. F. Arago, *Biographies of Distinguished Scientific Men,* 2nd Ser. (Boston, 1859), p. 160n.
3. H. Bence Jones (ed.), *The Life and Letters of Faraday* (London, 1870), II, p. 5.
4. F. Magie, *Source Book of Physics* (New York, 1935), p. 576. (Quotation of Sir William Crookes.)
5. Gerald Holton and Duane Roller, *Foundations of Modern Physical Science* (Reading, 1958), p. 692.

CHAPTER 16

1. *Phil, Trans.* (1852), 142, p. 417.

CHAPTER 17

1. W. F. Magie, *Source Book of Physics* (New York, 1935), p. 262.
2. *Ibid.*, 456.
3. *Ibid.*, 530.
4. Max Planck, *Scientific Autobiography* (London, 1950), p. 41.
5. *Ibid.*, p. 44.
6. *Ibid.*, p. 13.

CHAPTER 18

1. William Wordsworth, *The Prelude.*
2. *Punch,* Dec. 19, 1923.
3. *Science* (1962), 137, p. 731.
4. The New York *Times,* March 21, 1955.
5. William de Sitter, *Kosmos* (Cambridge, Mass., 1932), pp. 111-112.
6. Fred Hoyle, *Frontiers of Astronomy* (New York, 1955), pp. 354-355.

Further Reading

Chapter 1. J. H. Breasted (ed.), *The Edwin Smith Papyrus* (Chicago, 1930); O. Neugebauer, *The Exact Sciences in Antiquity,* 2nd ed. (Providence, R. I., 1957).

Chapter 2. Aristotle, *Physics* and *On the Heavens,* Marshall Clagett, *Greek Science in Antiquity,* Part I (New York, 1955); S. Sambursky, *The Physical World of the Greeks* (London, 1956).

Chapter 3. M. R. Cohen and I. E. Drabkin, *A Source Book in Greek Science* (New York, 1948); J. L. E. Dreyer, *A History of Astronomy from Thales to Kepler* (New York, Dover, 1953); Sir Thomas Heath, *A Manual of Greek Mathematics* (Oxford, 1931), *Greek Astronomy* (London, 1932).

Chapter 4. Marshall Clagett, *Greek Science in Antiquity,* Part II (New York, 1955); S. Sambursky, *The Physical World of Late Antiquity* (London, 1962).

Chapter 5. A. C. Crombie, *Medieval and Early Modern Science,* Vol. I (New York, Anchor, 1959); C. H.

Haskins, *The Renaissance of the Twelfth Century* (Cambridge, Mass., 1928), *Studies in the History of Mediaeval Science* (Cambridge, Mass., 1924); C. Singer, *From Magic to Science* (New York, Dover, 1958).

Chapter 6. Marshall Clagett, *The Science of Mechanics in the Middle Ages* (Madison, Wis., 1959); A. C. Crombie, *Medieval and Early Modern Science,* (New York, Anchor, 1959), *Robert Grosseteste and the Origins of Experimental Science, 1100–1700* (Oxford, 1953); Lynn White Jr. *Medieval Technology and Social Change* (Oxford, 1962).

Chapter 7. Marie Boas, *The Scientific Renaissance* (London, 1962); C. Singer, *History of Biology* (New York, 1950); T. H. White, *The Bestiary* (New York, Putnam, 1960).

Chapter 8. As Chapter 7, and Arturo Castiglioni, *History of Medicine* (New York, 1947); M. Foster, *History of Physiology* (Cambridge, Mass., 1901).

Chapter 9. E. J. Dijksterhuis, *The Mechanization of the World Picture* (Oxford, 1961); R. Dugas, *History of Mechanics* (London, 1957); R. Grant, *History of Physical Astronomy* (London, 1852); A. Rupert Hall, *The Scientific Revolution, 1500–1800* (London, 1962); A. Koyré, *From the Closed World to the Infinite Universe* (Baltimore, Md., 1957).

Chapter 10. As Chapter 9, and S. Drake, *Discoveries and Opinions of Galileo* (New York, Anchor, 1957); A. Rupert Hall, *Galileo to Newton* (London, 1963); L. T. More, *Isaac Newton* (New York, 1934); M. Ornstein,

The Role of Scientific Societies in the Seventeenth Century (Chicago, 1928).

Chapter 11. F. J. Cole, *History of Comparative Anatomy* (London, 1944); C. Dobell, *Antony van Leeuwenhoek and His "Little Animals"* (New York, 1958); J. Needham, *History of Embryology* (Cambridge, 1934).

Chapter 12. J. D. Bernal, *Science and Industry in the Nineteenth Century* (London, 1953); C. Singer, E. J. Holmyard, A. R. Hall, and T. I. Williams, *A History of Technology*, Vols. 4 and 5 (Oxford, 1958).

Chapter 13. L. Eiseley, *Darwin's Century* (New York, Anchor, 1958); C. C. Gillispie, *Genesis and Geology* (New York, Harper, 1959); J. C. Greene, *The Death of Adam* (New York, Mentor, 1961).

Chapter 14. A. Castiglioni, *History of Medicine* (New York, 1947); René J. Dubos, *Louis Pasteur* (Boston, 1950); Arthur Hughes, *History of Cytology* (New York, 1959); J. M. D. Olmsted, *François Magendie* (New York, 1944), *Claude Bernard* (New York, 1938); Richard S. Shryock, *Development of Modern Medicine* (New York, 1947).

Chapter 15. F. Cajori, *History of Physics* (New York, 1899); J. J. Thomson, *Recollections and Reflections* (London, 1936); John Tyndall, *Faraday as a Discoverer* (London, 1868).

Chapter 16. A. Findlay, *Hundred Years of Chemistry* (London, 1937); L. F. Haber, *Chemical Industry during the Nineteenth Century* (Oxford, 1958); H. M. Leicester, *The Historical Background of Chemistry* (New York, 1956); J. R. Partington, *Short History of Chemistry* (London, 1951).

Chapter 17. H. T. Pledge, *Science Since 1500* (New York, Harper's 1959); E. T. Whittaker, *From Euclid to Eddington* (Cambridge, 1949).

Chapter 18. A. M. Clerke, *History of Astronomy during the Nineteenth Century* (London, 1902); Philipp Frank, *Einstein* (London, 1948); E. Hiebert, *The Impact of Atomic Energy* (Newton, Kansas, 1961); R. Waterfield, *A Hundred Years of Astronomy* (New York, 1938).

[illegible] the Universal Genius (Oxford, 1958); H. M. [illegible], The Historical Development of Machines (New York, 1956); J. R. Partington, Short History of Chemistry (London, 1951).

Chapter 17. H. T. Pledge, Science Since 1500 (New York: Harper's 1959); E. T. Whittaker, From Euclid to Eddington (Cambridge, 1949).

Chapter 18. A. M. Clerke, History of Astronomy during the Nineteenth Century (London, 1902); Philipp Frank, Einstein (London, 1948); [illegible], The Conquest of Atomic Energy (New York, [illegible]); R. [illegible], A Hundred Years of Astronomy (New York, 1938).

Index